Signals and Communication Technology

This series is devoted to fundamentals and applications of modern methods of signal processing and cutting-edge communication technologies. The main topics are information and signal theory, acoustical signal processing, image processing and multimedia systems, mobile and wireless communications, and computer and communication networks. Volumes in the series address researchers in academia and industrial R&D departments. The series is application-oriented. The level of presentation of each individual volume, however, depends on the subject and can range from practical to scientific.

Indexing: All books in "Signals and Communication Technology" are indexed by Scopus and zbMATH

For general information about this book series, comments or suggestions, please contact Mary James at mary.james@springer.com or Ramesh Nath Premnath at ramesh.premnath@springer.com.

Sheikh Mohammad Idrees
Mariusz Nowostawski
Editors

Blockchain Transformations

Navigating the Decentralized Protocols Era

Editors
Sheikh Mohammad Idrees
Researcher DSE Lab
Department of Computer Science (IDI)
Norwegian University of Science
and Technology
Gjøvik, Norway

Mariusz Nowostawski
Associate professor
Norwegian University of Science
and Technology
Gjøvik, Norway

ISSN 1860-4862 ISSN 1860-4870 (electronic)
Signals and Communication Technology
ISBN 978-3-031-49595-3 ISBN 978-3-031-49593-9 (eBook)
https://doi.org/10.1007/978-3-031-49593-9

This Springer imprint is published by the registered company Springer Nature Switzerland AG
The registered company address is: Gewerbestrasse 11, 6330 Cham, Switzerland

Introduction

The book "Blockchain Transformations: Navigating the Decentralized Protocols Era" is a go-to guide, revealing how the amazing technology known as blockchain is reshaping various aspects of our lives—from education and health to banking and beyond. In a world that's always changing with technology, blockchain emerges as a real game-changer. This book is not just about the technology itself, but about the incredible transformations it brings to different aspects of our lives. This book will take you on a journey with blockchain through education, healthcare, digital identity, and more, revealing the potential for positive change. Each chapter is a window into the practical applications and real-world impacts of blockchain technology.

This book is for everyone—whether you're a curious learner, a tech enthusiast, or a professional seeking insights into the next wave of innovation. This book will take you on a trip to explore how this technology is making our world better. Let's dive in together into these chapters, explore the exciting world of decentralization, and discover new possibilities.

- **Empowering Education: Leveraging Blockchain for Secure Credentials and Lifelong Learning**
  Embark on a journey through the educational realm as we unveil the secure and lifelong learning opportunities facilitated by blockchain technology.
- **Utilization of Blockchain Technology in Artificial Intelligence–Based Healthcare Security**
  Explore the intersection of blockchain and artificial intelligence, unraveling the enhanced security measures in healthcare through innovative applications.
- **Decentralized Key Management for Digital Identity Wallets**
  Delve into the ecosphere of digital identity management as we navigate through decentralized key solutions in the realm of blockchain.
- **Towards Blockchain-Driven Solution for Remote Healthcare Service: An Analytical Study**
  Conduct a critical analysis of blockchain-driven solutions, particularly focusing on remote healthcare services and their transformative impact.

- **Smart Contract Vulnerabilities: Exploring the Technical and Economic Aspects**
  Uncover the technical and economic aspects surrounding smart contract vulnerabilities, offering insights into potential pitfalls and safeguards.
- **Modernizing Healthcare Data Management: A Fusion of Mobile Agents and Blockchain Technology**
  Witness the fusion of mobile agents and blockchain technology in revolutionizing healthcare data management, ensuring efficiency and security.
- **Machine Learning Approaches in Blockchain Technology-Based IoT Security: An Investigation on Current Developments and Open Challenges**
  Investigate the synergy between machine learning and blockchain in ensuring the security of the Internet of Things (IoT) and address current challenges.
- **Decentralized Identity Management Using Blockchain Technology: Challenges and Solutions**
  Navigate through the challenges and innovative solutions in decentralized identity management, highlighting the role of blockchain technology.
- **Reshaping the Education Sector of Manipur Through Blockchain**
  Witness the transformative impact of blockchain on the education sector, with a focus on reshaping the landscape in Manipur.
- **Exploring the Intersection of Entrepreneurship and Blockchain Technology: A Research Landscape Through R Studio and VOSviewer**
  Embark on a research journey exploring the intersection of entrepreneurship and blockchain, utilizing R Studio and VOS-viewer for a comprehensive landscape.
- **Transforming Educational Landscape with Blockchain Technology: Applications and Challenges**
  Uncover the applications and challenges associated with transforming the educational landscape through the integration of blockchain technology.
- **Verificate: Transforming Certificate Verification Using Blockchain Technology**
  Explore the innovative Verificate system, revolutionizing certificate verification through the seamless integration of blockchain technology.
- **Transforming Waste Management Practices Through Blockchain Innovations**
  Witness the positive environmental impact of blockchain innovations in transforming waste management practices.
- **Decentralized Technology and Blockchain in Healthcare Administration**
  Explore the decentralized technologies reshaping the landscape of healthcare administration, with a primary focus on blockchain applications.
- **Blockchain Technology Acceptance in Agribusiness Industry**
  Delve into the acceptance and integration of blockchain technology in the agribusiness industry, revolutionizing traditional practices.
- **Adoption of Blockchain Technology and Circular Economy Practices by SMEs**
  Analyze the adoption of blockchain technology and its alignment with circular economy practices among small- and medium-sized enterprises (SMEs).

Contents

Chapter 1
Empowering Education: Leveraging Blockchain for Secure Credentials and Lifelong Learning

Adil Marouan, Morad Badrani, Nabil Kannouf, and Abdelaziz Chetouani

1 Introduction

1.1 Background on Blockchain Technology and Its Applications Beyond Finance

Blockchain technology, initially developed for cryptocurrencies like Bitcoin, has garnered widespread attention due to its potential to revolutionize various industries beyond finance. Blockchain is a decentralized and distributed ledger that records transactions across multiple computers, ensuring transparency, immutability, and security. While finance was the initial domain where blockchain gained prominence, its applications have expanded to numerous sectors, including education [16, 22].

Blockchain technology provides several unique features [17] that make it suitable for applications beyond finance. One of the key features is decentralization, which means that no single entity has control over the entire blockchain network. Instead, the network participants, known as nodes, collectively maintain and validate the transactions and records. This decentralized nature eliminates the need for intermediaries, reducing costs and increasing efficiency.

Another crucial aspect of blockchain is immutability. Once a transaction or record is added to the blockchain, it cannot be altered or deleted. This feature ensures the integrity and trustworthiness of the data stored on the blockchain, making it highly resistant to tampering and fraud. Immutability is achieved through

A. Marouan (✉) · M. Badrani · A. Chetouani
LaMAO Laboratory, ORAS team, ENCG, Mohammed First University, Oujda, Morocco
e-mail: adil.marouan@ump.ac.ma; m.badrani@ump.ac.ma; a.chetouani@ump.ac.ma

N. Kannouf
LSA laboratory, SOVAI Team, ENSA, Abdelmalek Essaadi University, Alhoceima, Morocco

S. M. Idrees, M. Nowostawski (eds.), *Blockchain Transformations*, Signals and Communication Technology, https://doi.org/10.1007/978-3-031-49593-9_1

cryptographic techniques and consensus algorithms, ensuring that all network participants agree on the validity of transactions [4, 24].

Furthermore, blockchain technology offers enhanced security. Data stored on the blockchain is encrypted and linked to previous transactions, creating a chain of blocks that are nearly impossible to manipulate without consensus from the network. Additionally, the decentralized nature of the blockchain reduces the risk of a single point of failure and makes it more resilient against cyberattacks.

Beyond finance [2], blockchain technology has the potential to transform the field of education. By leveraging its unique features, blockchain can address various challenges related to data privacy, security, verification, and accessibility in education.

1.2 Importance of Exploring Blockchain Technology in Education

Blockchain technology has emerged as a transformative force across various industries, and its potential in the field of education is gaining significant attention. Blockchain, often associated with cryptocurrencies like Bitcoin, is essentially a decentralized and transparent digital ledger that records and verifies transactions. However, its application extends far beyond financial systems, offering unique advantages that can revolutionize the education sector (Fig. 1.1).

This chapter aims to explore the potential of BCT in revolutionizing education. It addresses key issues related to data privacy, security, verification, and accessibility within the education system.

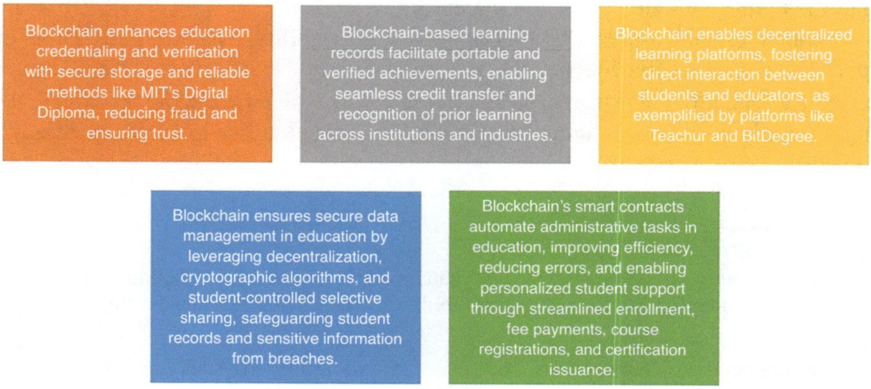

Blockchain enhances education credentialing and verification with secure storage and reliable methods like MIT's Digital Diploma, reducing fraud and ensuring trust.

Blockchain-based learning records facilitate portable and verified achievements, enabling seamless credit transfer and recognition of prior learning across institutions and industries.

Blockchain enables decentralized learning platforms, fostering direct interaction between students and educators, as exemplified by platforms like Teachur and BitDegree.

Blockchain ensures secure data management in education by leveraging decentralization, cryptographic algorithms, and student-controlled selective sharing, safeguarding student records and sensitive information from breaches.

Blockchain's smart contracts automate administrative tasks in education, improving efficiency, reducing errors, and enabling personalized student support through streamlined enrollment, fee payments, course registrations, and certification issuance.

Fig. 1.1 Applications of BCT in education

2 Blockchain Technology and Education

2.1 Unique Features of Blockchain That Can Address Educational Challenges

Blockchain technology possesses several unique features that have the potential to address various challenges faced in the field of education. This section will explore some of these features and their potential applications in addressing educational challenges.

(a) Data Integrity and Security: Blockchain's inherent design ensures data integrity and security. The immutability of data stored on the blockchain makes it highly resistant to tampering or unauthorized modifications. This feature can be leveraged to address challenges related to student record management, certificate authentication, and academic credential verification [3], storing educational records and credentials on the blockchain, educational institutions can maintain a reliable and tamper-proof repository of student achievements, ensuring the authenticity and security of educational data [21].
(b) Transparent and Trustworthy System: Blockchain's transparency and decentralized nature create a trustworthy system for educational transactions. Smart contracts, self-executing agreements built on blockchain, can facilitate transparent and automated processes in various areas, such as student enrollment, course registration, and financial transactions. These smart contracts can streamline administrative processes, reduce fraud, and enhance trust among stakeholders [2].
(c) Portable and Lifelong Learning Records [14]: Blockchain technology enables the creation of portable and interoperable learning records. Students can have ownership and control over their educational achievements, which can be securely stored on the blockchain. This feature allows for the seamless transfer of learning records between educational institutions, supporting lifelong learning and facilitating credential recognition.
(d) Microcredentialing and Personalized Learning: Blockchain can enable the issuance and management of microcredentials, which are digital badges representing specific skills or competencies. Microcredentials can be verified and shared securely, allowing individuals to demonstrate their skills beyond traditional degrees or certifications. This supports personalized learning pathways, enabling learners to showcase their diverse skills and achievements.
(e) Enhanced Collaboration and Content Sharing: Blockchain technology can facilitate decentralized and peer-to-peer collaboration among learners and educators. Blockchain-based platforms can provide secure environments for sharing educational resources, fostering collaboration, and incentivizing contributions through tokens or rewards. This decentralized approach promotes the creation and sharing of open educational resources, encouraging innovation and knowledge exchange.

By leveraging the unique features of blockchain, educational institutions can overcome challenges related to data security, transparency, portability of records, and collaboration. However, it is important to carefully consider the implementation of blockchain in education and address technical, regulatory, and ethical considerations to maximize the potential benefits.

2.2 Potential Benefits of Implementing Blockchain in Education

Implementing blockchain technology in education holds the potential to bring about several significant benefits. This section will explore some of the potential advantages that blockchain can offer to the field of education.

Enhanced Data Security and Privacy: Blockchain ensures data security and privacy by decentralizing and making it tamper-resistant, enhancing the protection of educational records and sensitive student information [3]. This can help protect against data breaches and ensure the integrity of academic records.

Improved Verification and Credentialing: Blockchain revolutionizes credential verification by creating a tamper-proof repository of academic records, enabling easy and authentic verification for employers and institutions, reducing reliance on paper-based methods [14]. This streamlined and efficient verification process can help address issues related to credential fraud and enhance trust in the educational system.

Increased Transparency and Accountability: Blockchain's transparency and auditability promote accountability in education by recording transactions and ensuring transparent processes, preventing fraud and fostering integrity in institutions [2]. Additionally, the decentralized nature of blockchain can foster trust among stakeholders, as all participants have access to the same verified information.

Facilitated Micropayments and Royalties: Blockchain technology enables the use of smart contracts, which are self-executing contracts with predefined rules and conditions. Smart contracts can facilitate micropayments and royalties for educational content creators, such as authors, instructors, or developers of educational resources. Through blockchain-based platforms, creators can receive fair compensation for their work, fostering innovation and encouraging the production of high-quality educational materials [14].

Streamlined Administrative Processes: Blockchain has the potential to streamline administrative processes in the education sector. By leveraging smart contracts, tasks such as student enrollment, course registration, and financial transactions can be automated and executed with increased efficiency. This can reduce administrative burdens, minimize errors, and free up valuable resources for educational institutions.

Open and Collaborative Educational Ecosystem: Blockchain technology can facilitate the creation of an open and collaborative educational ecosystem. It can enable the sharing and verification of educational resources, fostering collaboration among educators and learners. Blockchain-based platforms can provide secure environments for the creation, sharing, and adaptation of open educational resources, ensuring attribution, and incentivizing contributions [14].

2.3 Examples of Universities and Educational Institutions Implementing Blockchain

These examples showcase how universities around the world have started to integrate blockchain technology into their educational processes, ranging from issuing digital credentials to conducting research and offering specialized courses. The adoption of blockchain in education is still evolving, and more institutions are likely to explore its potential in the future:

- **MIT Media Lab (Massachusetts Institute of Technology)**: The MIT Media Lab has developed a blockchain-based platform called "Blockcerts" for issuing and verifying digital credentials. This platform allows students to store and share their academic achievements securely using blockchain technology.
- **Stanford University:** Stanford has conducted research on using blockchain to secure and streamline academic transcripts. The university explored how blockchain can enhance data security and reduce administrative burdens related to transcript management.
- **University of Sydney**: The University of Sydney has explored blockchain technology to create a platform for students to store and share their academic credentials securely. This initiative aims to simplify the verification process for both students and employers.
- **King Abdullah University of Science and Technology (KAUST)**: KAUST in Saudi Arabia has partnered with IBM to explore the use of blockchain technology in various educational and research contexts.
- **Mohammed First University**: Researchers in Moroccan universities are currently working on utilizing blockchain technology in electronic voting.

3 Verifiable Digital Credentials

3.1 The Need for Secure and Verifiable Digital Credentials in Education

In the digital era, the traditional methods of issuing and verifying educational credentials are facing challenges in terms of security, portability, and efficiency. As a result, there is a growing need for secure and verifiable digital credentials in the field of education. Some of the reasons why secure and verifiable digital credentials are crucial in education and provide supporting references are mitigating credential fraud. Credential fraud, including the fabrication or alteration of academic achievements, is a significant concern in the education sector. Traditional paper-based credentials are susceptible to forgery and tampering, making it difficult to trust the authenticity of qualifications. Verifiable digital credentials, on the other hand, utilize cryptographic techniques to ensure the integrity and immutability of the credential data [14]. By implementing secure digital credentialing systems, educational institutions can mitigate the risk of credential fraud and enhance trust in the qualifications of their students.

The second reason is enabling lifelong learning. In today's rapidly evolving job market, lifelong learning has become essential for individuals to adapt and upskill. However, the recognition of informal and non-traditional learning experiences poses a challenge. Verifiable digital credentials can address this challenge by providing a mechanism to capture and represent various forms of learning, including online courses, workshops, and work-based learning [3]. These digital credentials can be easily updated, stacked, and shared, enabling individuals to showcase their continuous learning journey.

3.2 *Exploring the Use of Blockchain for Storing and Authenticating Credentials*

Blockchain technology has garnered significant attention in recent years, not only for its association with cryptocurrencies but also for its potential to revolutionize various industries. One area where blockchain shows great promise is in the storage and authentication of credentials. We use blockchain for the purposes on Fig. 1.2 and highlight its benefits and challenges.

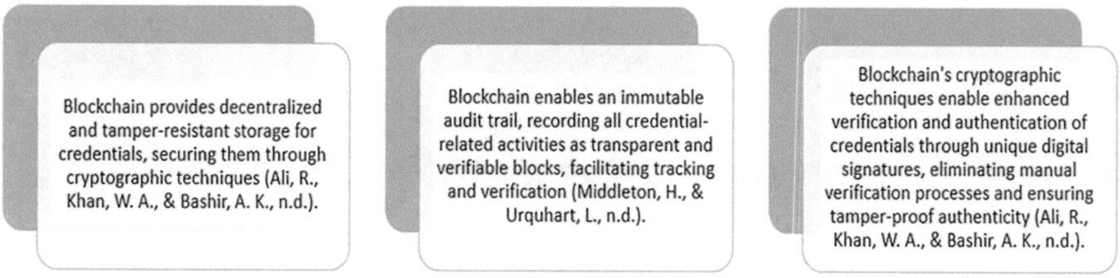

Fig. 1.2 The use of BCT in storage and authentication of credentials

4 Security and Privacy of Student Data

4.1 Current Challenges in Protecting Student Data

In the digital age, educational institutions and organizations are increasingly relying on technology to manage and store student data. This shift has raised concerns about the security and privacy of student information. Safeguarding student data is of paramount importance to protect sensitive personal information and ensure trust within educational systems. Blockchain technology has emerged as a potential solution to enhance the security and privacy of student data. Fig. 1.3 explores the key considerations and benefits associated with using blockchain in the context of student data security and privacy.

4.2 Leveraging Blockchain for Secure and Private Data Storage

Blockchain technology has gained significant attention due to its potential for secure and transparent data management across various industries. One area where blockchain shows promise is in secure and private data storage. By utilizing its decentralized nature, immutability, and cryptographic algorithms, blockchain can provide robust solutions for protecting sensitive data from unauthorized access, tampering, and breaches. In this section, we will explore the key features of blockchain that make it suitable for secure and private data storage, as well as discuss some notable references in this field.

Decentralization and Data Redundancy: Blockchain's decentralized architecture eliminates the need for a central authority, such as a server or database, to store and manage data. Instead, data is distributed across a network of nodes, ensuring

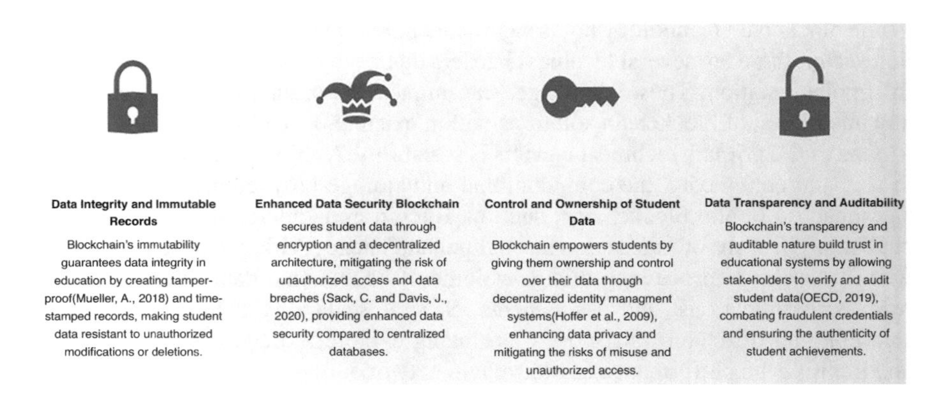

Fig. 1.3 Benefit of BCT for student data privacy

redundancy and fault tolerance. This decentralized approach reduces the risk of a single point of failure, making it challenging for hackers to compromise the data. Additionally, storing multiple copies of data across the network enhances data availability and integrity.

Immutability and Data Integrity: Blockchain achieves data immutability through the use of cryptographic hashing algorithms [16] and consensus mechanisms. Each data block is cryptographically linked to the previous block, forming a chain of blocks. Once a block is added to the chain, it becomes computationally infeasible to alter or delete its contents without invalidating the entire chain. This feature ensures data integrity, as any unauthorized modification attempts can be easily detected.

Encryption and Access Control: Blockchain technology can leverage advanced encryption techniques [7] to protect the confidentiality of stored data. By encrypting data before storing it on the blockchain, sensitive information remains secure even if the blockchain's content is publicly accessible. Additionally, access control mechanisms, such as public-private key pairs, can be implemented to grant authorized parties the ability to decrypt and access specific data.

Smart Contracts and Data Management: Smart contracts are self-executing agreements with predefined rules and conditions stored on the blockchain. They provide an additional layer of security and automation for data storage and access. Smart contracts can enforce access controls, verify data integrity, and execute predefined actions based on specified conditions. By leveraging smart contracts, blockchain-based data storage systems can ensure secure and reliable data management [19, 23].

5 Challenges and Limitations

5.1 Technical Barriers in Implementing Blockchain in Education

While blockchain technology holds significant potential for transforming the education sector, there are several technical barriers that need to be addressed for successful implementation. These challenges can impact the scalability, interoperability, and integration of blockchain solutions within existing educational systems.

One of the primary technical barriers is scalability. As blockchain networks grow in size and complexity, the computational and storage [10] requirements increase significantly. Public blockchains, such as Bitcoin and Ethereum, face scalability challenges in terms of transaction throughput and latency. This can pose a hindrance when it comes to processing a large volume of educational data, such as student records, certifications, and assessments. Several scalability solutions, including sharding and off-chain transactions, are being explored to address this challenge and improve the performance of blockchain networks.

Interoperability [20] is another technical hurdle in implementing blockchain in education. Educational institutions typically use a wide range of systems and platforms to manage student data, learning management systems, and other educational resources. Achieving seamless integration between blockchain and these existing systems is crucial to enable effective data exchange and interoperability. Efforts are underway to develop standardized protocols, such as the InterPlanetary File System (IPFS) and the W3C Verifiable Credentials, to facilitate interoperability between different blockchain networks and educational platforms.

Security and privacy considerations also present technical challenges. While blockchain technology itself provides strong security through cryptography and immutability, securing the underlying infrastructure, such as storage, communication channels, and user identities [13], is critical. Protecting sensitive student data, such as personally identifiable information (PII), from unauthorized access or data breaches requires robust security measures. Additionally, ensuring data privacy while leveraging the transparency and traceability of blockchain technology requires careful design and implementation.

Furthermore, user experience and accessibility are important aspects to consider. The usability of blockchain applications and interfaces should be intuitive and user-friendly, ensuring that educators, students, and administrators can easily interact with the blockchain system. The technical complexity [12] associated with blockchain technology should be abstracted to provide a seamless user experience, enabling widespread adoption within educational settings.

Addressing these technical barriers requires collaboration between educational institutions, technology providers, and blockchain experts. Research and development efforts are underway to create scalable blockchain solutions, enhance interoperability, strengthen security measures, and improve user experience in educational blockchain applications.

5.2 Regulatory Considerations and Compliance Issues

Implementing blockchain technology in the education sector requires careful attention to regulatory considerations and compliance issues. As blockchain solutions involve the storage and management of sensitive student data, educational institutions must navigate legal and regulatory frameworks to ensure compliance with relevant laws and regulations.

One significant regulatory consideration is data protection and privacy. Many jurisdictions have stringent data protection laws, such as the European Union's General Data Protection Regulation (GDPR) and the California Consumer Privacy Act (CCPA). These regulations impose strict requirements for the collection, storage, and processing of personal data, including student information. Educational institutions [5] adopting blockchain must ensure that their blockchain implementations adhere to these regulations, providing appropriate consent mechanisms, data minimization, and secure data handling practices.

In addition to data privacy, compliance with academic standards and accreditation requirements is essential. Educational institutions must ensure that blockchain-based systems for storing and verifying credentials meet the standards set by relevant accrediting bodies. This involves establishing trust in the blockchain system and ensuring the accuracy and integrity of the stored credentials. Collaborating with accrediting bodies and regulatory agencies can help ensure that blockchain implementations in education comply with the necessary standards and regulations [6].

Furthermore, intellectual property rights and copyright considerations should not be overlooked. Blockchain technology [11] enables the transparent sharing and distribution of educational content, raising questions about ownership and licensing rights. Educational institutions [15] must navigate copyright laws and establish clear guidelines regarding the use and distribution of educational materials on the blockchain. Collaborations with content creators, licensing agencies, and legal experts can help address these regulatory challenges effectively.

It is important to recognize that regulatory frameworks surrounding blockchain technology in education are still evolving. As such, educational institutions should actively monitor updates and engage in discussions with policymakers and regulatory authorities to shape the regulatory landscape and ensure compliance with emerging requirements.

5.3 Scalability Challenges and Potential Solutions

Scalability is a significant challenge when implementing blockchain technology, as it involves handling a large volume of transactions and data within a network. This challenge becomes particularly crucial in sectors like education, where the storage and processing of extensive student records, certificates, and assessments are involved. Addressing scalability concerns is crucial to ensure that blockchain-based educational systems can handle increased transactional demands and accommodate growing user bases [1].

One of the primary scalability challenges in blockchain technology is transaction throughput. Public blockchains, such as Bitcoin and Ethereum, have limitations in terms of the number of transactions they can process per second. For instance, Bitcoin can handle around 7 transactions per second, while Ethereum's throughput [9] is higher but still limited. This limitation can result in delays and increased transaction fees, impeding the seamless flow of data and transactions within an educational blockchain ecosystem.

To overcome these challenges, several potential solutions are being explored. One approach is the implementation of off-chain transactions or layer-2 scaling solutions. These solutions involve conducting certain transactions or computations off the main blockchain, reducing the load on the main network. Off-chain transactions can be settled periodically on the main blockchain, maintaining the security and immutability of the underlying data.

Another solution is the concept of sharding, which involves dividing the blockchain network into smaller, interconnected sub-networks known as shards. Sharding allows for parallel processing of transactions and data across multiple shards, significantly increasing the overall capacity and transaction throughput of the blockchain network. By distributing the workload across shards, scalability can be improved while maintaining the security and decentralization aspects of the blockchain.

Additionally, advancements in consensus algorithms offer potential scalability improvements. Traditional proof-of-work algorithms, while secure, are computationally intensive and limit scalability. Alternative consensus mechanisms [18] like proof-of-stake, proof-of-authority, or delegated proof-of-stake can provide higher transaction throughput and lower energy consumption, enabling greater scalability for blockchain networks.

Furthermore, advancements in infrastructure, such as high-performance hardware and optimized software, can contribute to improved scalability. Increasing network bandwidth, storage capabilities, and computational power can help handle larger volumes of data and transactions within a blockchain system.

It is important to note that scalability solutions for blockchain technology are still evolving, and their effectiveness and applicability may vary depending on the specific use case and requirements of educational blockchain implementations. Ongoing research and development efforts continue to explore innovative approaches to address scalability challenges in blockchain technology.

6 Future Research and Experimentation

6.1 Importance of Continued Research and Experimentation

Continued research and experimentation play a vital role in unlocking the full potential of blockchain technology in the education sector. As blockchain is a relatively new and evolving technology, there are still many aspects to explore, refine, and optimize. Investing in research and experimentation allows educational institutions to stay at the forefront of innovation and harness the transformative power of blockchain in education.

One key importance of continued research is to address technical challenges and limitations associated with blockchain implementation. Researchers can focus on scalability, interoperability, security, and privacy issues to develop solutions that overcome these hurdles. Through experimentation, researchers can test new consensus algorithms, explore novel approaches to data storage and management, and propose frameworks for secure and efficient integration of blockchain with existing educational systems.

Moreover, research efforts can help identify and understand the potential benefits and impacts of blockchain technology in education. Studies can evaluate the effectiveness of blockchain-based solutions in improving processes such as credential verification, lifelong learning tracking, and decentralized educational content sharing. By examining real-world use cases and conducting empirical studies, researchers can provide insights into the value proposition of blockchain and guide its adoption in education.

Continued research also facilitates collaboration among academia, industry, and policymakers. By fostering interdisciplinary dialogue, researchers can contribute to the development of regulatory frameworks, standards, and best practices for blockchain implementation in education. This collaboration ensures that blockchain solutions align with legal and ethical considerations while addressing the specific needs and challenges of educational institutions.

Furthermore, research and experimentation can drive innovation in the design and implementation of user-friendly blockchain interfaces and educational applications. By focusing on user experience, researchers can make blockchain technology more accessible and intuitive for educators, students, and administrators. This includes simplifying complex concepts, improving the usability of blockchain interfaces, and designing intuitive educational platforms that leverage blockchain's benefits.

In summary, continued research and experimentation are essential for advancing the adoption and effectiveness of blockchain technology in education. It enables the development of scalable, secure, and user-friendly solutions; informs policy and standards; and explores the full potential of blockchain in transforming educational systems.

6.2 Exploring and Refining the Potential of Blockchain Technology in Education

Blockchain technology has gained significant attention in various industries due to its potential to enhance transparency, security, and efficiency. In recent years, the education sector has also recognized the value of blockchain in revolutionizing traditional processes and transforming the way educational credentials are managed and verified. This section explores the potential applications of blockchain technology in education and highlights the need for further refinement and exploration.

One of the key areas where blockchain (L. Uden, M. Sinclair, Y.-H. Tao, D. Liberona, & R. M. A. Pinto (Eds.) Anderson, S., 2016) can make a significant impact is in the verification and authentication of educational certificates and credentials. With the current system heavily reliant on paper-based records and manual verification processes, the potential for fraud and misrepresentation is high. Blockchain technology can offer a decentralized and immutable ledger, ensuring the authenticity of educational records. Employers and academic institutions can easily verify the credentials of applicants, saving time and resources.

Moreover, blockchain-based systems can facilitate the secure transfer and storage of academic records, enabling students to have complete ownership and control over their educational data. This empowers learners to share their achievements with potential employers, institutions, or other stakeholders, eliminating the need for intermediaries and enhancing data privacy.

Furthermore, blockchain-based microcredentialing systems have the potential to revolutionize lifelong learning and skills development. These systems can provide learners with the ability to earn and store digital badges or tokens for completing specific courses or acquiring specific skills. These digital credentials can be easily verified and shared, enabling employers and educational institutions to assess an individual's competency and skill set accurately [8].

Despite the promising potential of blockchain in education, further exploration and refinement are necessary to address challenges and ensure successful implementation. Issues such as scalability, interoperability, and standardization need to be carefully considered. Scalability challenges arise due to the decentralized nature of blockchain networks, as they require significant computational power and storage capacity. Innovative solutions such as off-chain scaling techniques and layer-two protocols can help overcome these challenges.

Moreover, regulatory considerations and compliance issues play a crucial role in the adoption of blockchain technology in education. Compliance with data protection and privacy laws, as well as addressing concerns related to data ownership and control, is essential. Collaboration between educational institutions, policymakers, and blockchain technology providers is necessary to establish a robust regulatory framework that ensures data security and privacy while promoting innovation.

7 Conclusion

This chapter has highlighted the key points and findings regarding the potential of blockchain technology in education. It has emphasized the benefits of blockchain in verifying credentials, enabling secure record-keeping, and revolutionizing micro-credentialing. However, challenges such as scalability need to be addressed for successful implementation. Overall, blockchain has the potential to transform education by enhancing transparency, security, and learner empowerment.

In summary, the implementation of blockchain in education presents various implications and challenges. While it holds promise for verifying credentials, ensuring data security, and enabling lifelong learning, scalability and regulatory considerations need to be carefully addressed. Overcoming these challenges will be crucial to fully harnessing the potential benefits of blockchain technology in education.

In final thoughts, blockchain technology has the transformative potential to revolutionize education. Its ability to enhance transparency, security, and ownership of educational records can empower learners and streamline processes. However, successful implementation requires addressing scalability challenges, regulatory considerations, and fostering collaboration among stakeholders. With careful refinement and exploration, blockchain has the capacity to reshape education, making it more accessible, efficient, and learner-centric.

References

1. Al-Bahri, M., Al-Bahri, A., Al-Khalifa, H., & Al-Hassan. (2019). Survey and evaluation of scalability of blockchain consensus algorithms. In *2019 International Symposium on Networks, Computers and Communications (ISNCC)*, pp. 1–6.
2. Al-Busaidi, A. A., & Mayhew, L. (s.d.). *Blockchain technology: Implementation challenges in global supply chains. 157*, 120079.
3. Ali, R., Khan, W. A., & Bashir, A. K. (s.d.). *Blockchain-based decentralized system for reliable student data management. 142*, 103647.
4. Antonopoulos, A. M. (2014). *Mastering Bitcoin: Unlocking digital cryptocurrencies*. O'Reilly Media, Inc.
5. Azzi, N., & Jarju, B. S. (s.d.). *Privacy preserving blockchain-based certificate verification: a review*. 94–107.
6. Bartolucci, F., & Donnelly, K. (2019). Blockchain in education: Examining potential, pitfalls, and perspectives. In *Proceedings of the 52nd Hawaii international conference on system sciences.*
7. Benet, J. (2014). Ipfs-content addressed, versioned, p2p file system. *arXiv preprint arXiv:1407.3561.*
8. Choudhury, S. R., & Abraham, A. (s.d.). A blockchain future for higher education? In *Proceedings of the 2018 international conference on data science and computational intelligence*, pp. 44–49.
9. Croman, K., Decker, C., Eyal, I., Gencer, A. E., Juels, A., Kosba, A., … & Song, D. (2016). On scaling decentralized blockchains. In *International conference on financial cryptography and data security*, pp. 106–125.
10. Gao, F., & Xu, X. (s.d.). *Privacy preservation in blockchain: Research challenges and opportunities*. 22218–22236.
11. Griggs, K., & Jackson, E. (s.d.). Blockchain in education. *EdTech, 15*(5), 10–14.
12. Gupta, R. K., Patel, N., & Nandini, K. (2021). Blockchain and data protection regulations: Challenges, opportunities and future prospects. *Computers & Security, 108.*
13. Janssen, B., & Debruyne, C. (2020). Blockchain in education: Risks and opportunities. *Frontiers in Blockchain*, 3–15.
14. Khan, I. A., & Iqbal, W. (s.d.). Blockchain technology in education: Opportunities, challenges, and solutions. *IEEE Access, 9*, 107211–107230.
15. Kshetri, N. (2018). 1 Blockchain's roles in meeting key supply chain management objectives. *International Journal of Information Management, 39*, 80–89.
16. Uden, L., Sinclair, M., Tao, Y. -H., Liberona, D., Pinto, R. M. A., & Anderson, S. (Eds.) (2016). *Blockchain technology and education. Learning Technology for Education in Cloud—The Changing Face of Education : 5th International Workshop, LTEC 2016* (Vol. 738, p. 214–222). https://doi.org/10.1007/978-3-319-71940-5_21
17. Middleton, H., & Urquhart, L. (s.d.). Blockchain in education. *International Journal of Educational Technology in Higher Education, 15*, 1–24.
18. Nakamoto, S. (2008). Bitcoin: A peer-to-peer electronic cash system. *Decentralized Business Review, 21260.*
19. Oosterhof, H., & Snijders, N. (s.d.). GDPR and blockchain: Friends or foes? *International Data Privacy Law, 10*(3), 227–246.
20. Swan, M. (2015). *Blockchain: Blueprint for a new economy*. O'Reilly Media, Inc.
21. Tapscott, D., & Tapscott, A. (2016). *Blockchain revolution: How the technology behind bitcoin is changing money, business, and the world*. Penguin.
22. Tschorsch, F., & Scheuermann, B. (2016). Bitcoin and beyond: A technical survey on decentralized digital currencies. *IEEE Communications Surveys & Tutorials, 18*(3), 2084–2123.
23. Wilkinson, S., Boshevski, T., Brandoff, J., & Buterin, V. (2014). *Storj a peer-to-peer cloud storage network*. Citeseer.
24. Yin, J., Qin, Z., & Wen, Q. (2020). Blockchain technology in education: Recent advances and future prospects. *Computers & Education, 103723*, 145.
25. Yli-Huumo, J., Ko, D., Choi, S., Park, S., & Smolander, K. (2016). Where is current research on blockchain technology?—A systematic review. *PLoS One, 11*(10), e0163477.

Chapter 2
Utilization of Blockchain Technology in Artificial Intelligence–Based Healthcare Security

Pranay Shah, Sushruta Mishra, and Angelia Melani Adrian

1 Introduction

The healthcare industry is in dire need of reform, from infectious diseases to cancer to radiography. There are various ways to use technology to deliver more precise, trustworthy, and efficient solutions. Artificial intelligence is the technology used to carry out works that would typically need human insights. Machine learning is a branch of AI that enables us to improve the AI algorithm by utilizing large amounts of data collected dynamically. AI is capable of comprehending and interpreting language, analyzing audio, recognizing things, and finding patterns to perform various kinds of tasks. This chapter demonstrates different perspectives of artificial intelligence in audio, video, and text and the challenges it faces in healthcare. Artificial intelligence includes natural language processing (NLP), machine learning for healthcare imaging, and acoustic AI. Natural language processing assists to enable computers to grasp texts and languages like that of humans. Once completed, computer systems interpret, sum up, and synthesize precise text and language from the given data. The healthcare sector produces a considerable quantity of written information, such as clinical reports, lab results, handwritten notes, admission and discharge records, and others, as illustrated in Fig. 2.1. Interpreting and managing such an enormous volume of data manually would be extremely challenging for healthcare professionals. The three main tasks that can be driven by NLP are opinion mining, information classification, and extracting significant facts from the text. NLP assists by analyzing and transforming these expanding data sets into a

P. Shah · S. Mishra (✉)
Kalinga Institute of Industrial Technology, Deemed to be University,
Bhubaneswar, Odisha, India
e-mail: sushruta.mishrafcs@kiit.ac.in

A. M. Adrian
Universitas Katolik De La Salle Manado, Manado, Indonesia

S. M. Idrees, M. Nowostawski (eds.), *Blockchain Transformations*, Signals and Communication Technology, https://doi.org/10.1007/978-3-031-49593-9_2

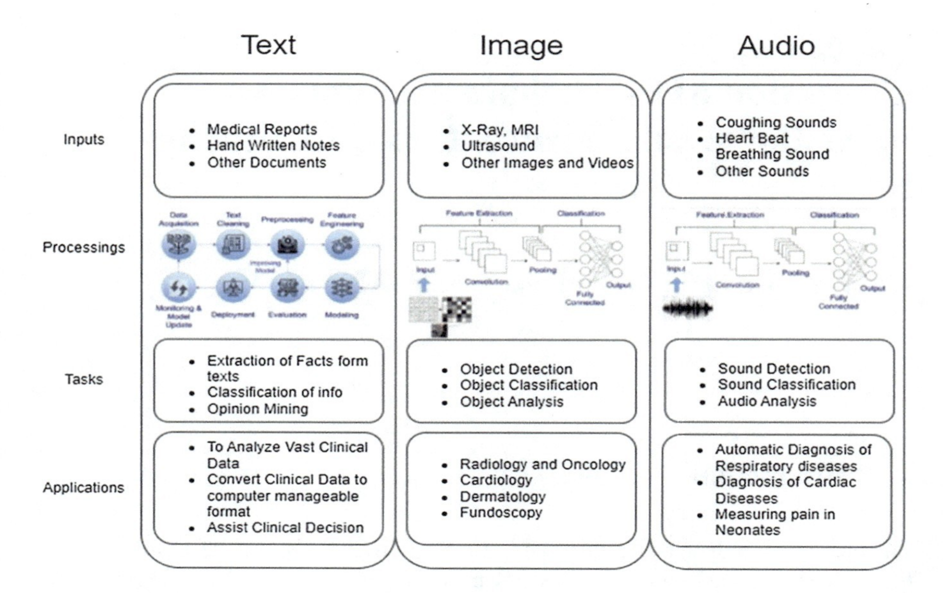

Fig. 2.1 Illustration of machine intelligence in healthcare

computer-manageable format. Additionally, it can support professional judgment; pinpoint vulnerable patients; and categorize diseases, syndromes, symptoms, and disorders. Similarly, machine learning for healthcare imaging is done through computer vision. Computer vision focuses on teaching computer systems to imitate human sight and analyze and interpret the things around them. Using artificial intelligence algorithms that evaluate images, computer vision does this. Medical records such as X-ray reports, CT scan, MRI scan, images of ultrasound, and videos play a vital role in a patient's diagnosis. Computer vision can improve the monitoring of patients, diagnose automatically, and generate lab reports automatically through various techniques such as object identification, categorization, location, and analysis from images or videos. It can encourage the development of a variety of applications that could save patients' lives in the fields of dermatology, oncology, cardiology, radiology, and fundoscopy. Similarly, as shown in the audio part of Fig. 2.1, different patterns of sounds of breathing, beating of heart, wheezing, crying, coughing, and so on, have a significant part in the diagnosis of various respiratory, pulmonary, and cardiac-related diseases. By examining noises, categorizing them, and evaluating them along the spectrum of audio, AI helps automate these diagnoses. Modern machine learning algorithms are available for processing audio signals, useful in the healthcare sector. For AI models to be more widely used in the healthcare sector, they must overcome several standards and problems, as shown in Fig. 2.1. First, when trained on sufficiently enough datasets, AI models can function precisely. Hence, one of the main difficulties is finding large, reliable, and trustworthy datasets for training. Second, it might be feasible if we chose to compile information from several sources and protect data from confidentiality violations and security threats. Third, because AI models are inherently opaque, it can be challenging to spot biased algorithms. In order to address the issue of mistrust in the learned model, it is necessary to have a record of the prediction or classification that arises from a particular healthcare input. If the incorrect course of action is taken based on AI conclusions, human lives are at risk. Fourth, there should be safe resource sharing to combat the threat posed by rogue devices. Finally, difficulties with information privacy exist when researchers and clinicians share their knowledge. As a result, there needs to be a tested plan to get through these obstacles before AI takes over the healthcare sector.

Learning and exploring through data increases the awareness and efficiency of AI-based healthcare in terms of accurate diagnosis and treatment planning while posing issues with anonymity, data governance, and the ability to generate income from sensitive patient data. So, it is necessary to make certain that the quality of the final therapy is good and that the patients quickly get the treatment required to manage their acute or chronic illnesses. The use of AI and ML can significantly enhance healthcare. However, the potential for adversarial attacks on natural language processing, AI healthcare imaging, and acoustic AI poses a constraint on their widespread use. In highly sensitive application domains like healthcare, these assaults cannot be tolerated. Considering the threats in these fields, blockchain can defend against these adversarial assaults. The intersection of blockchain technology and healthcare based on AI has the capability to completely change security and privacy. This review shows the possible use of AI-based blockchain integration in the sector of healthcare. In order to strengthen AI-based healthcare systems, this research relates the security measures using blockchain for the adversarial threats in NLP, healthcare imaging using ML (computer vision), and acoustic AI.

The main contributions of this chapter are as follows:

- A brief introduction to existing AI technologies in healthcare is discussed and the role of blockchain in clinical security is highlighted.
- A succinct overview of the existing advanced security based technologies used in the healthcare system is discussed.
- Various security threats, and adversarial attacks on AI technologies that are used in healthcare (NLP, medical imaging using machine learning, acoustic AI) are presented along with modern blockchain solutions.

2 Outline of Advanced Technologies Used in Healthcare

Advanced technologies have been integrated into the healthcare system nowadays to assist with patient monitoring, diagnosis, treatment, research, decision-making, hospital management, and so on. Some of these are the Internet of Things, blockchain, cloud computing, and artificial intelligence. They also provide automation, intelligence, security, and a low-cost computational ecology, which forms the base and enhances the functionality of the established healthcare system. The key enabling technologies and their main attributes used are shown in Fig. 2.2. Also, Table 2.1 displays a comparison of the numerous services provided by contemporary technologies including blockchain, IoT, AI, and cloud computing.

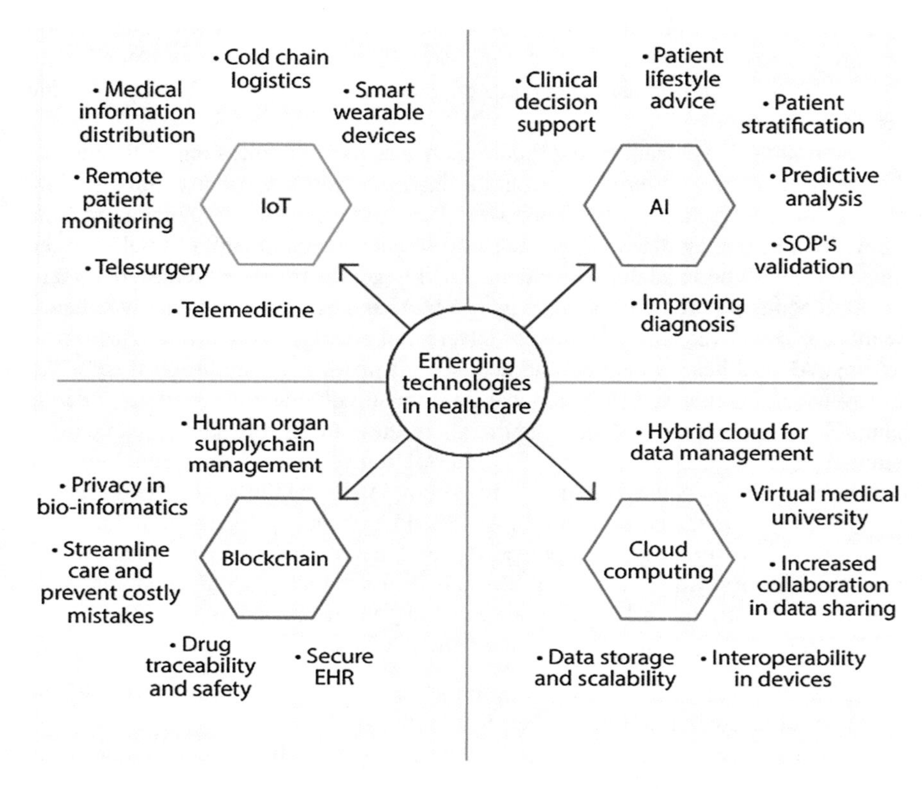

Fig. 2.2 Technologies used in advanced healthcare systems and their features

Table 2.1 Comparison of different services provided by advanced technologies

Modern technologies	Advantages	Challenges
Blockchain	Security, immutable, decentralized, Trustable, transparent	Requires more bandwidth, relatively expensive and complex than existing databases
Artificial intelligence	Compatible with different platforms. Versatile, reliable, efficient	Complex and difficult to design, expensive and difficult for deployment
Cloud computing	Efficiency, more capacity, ease of data storage, flexible	Difficult to manage, security threats, privacy concerns, high cost of communication
Internet of things	Low latency, portable, availability, efficient algorithm	Poor computation, privacy concerns, traffic

3 Blockchain-Enabled Technology in Healthcare

Blockchain technology was initially created to provide support for the use of cryptocurrency. However, recently it has been applied to different fields, achieving exceptional security [1]. At present, the healthcare industry has begun to integrate blockchain into several aspects of its operations. Its attributes, such as decentralized exchanges, micro-transactions, smart contracts, and consensus mechanisms, can help safeguard the confidentiality of patient data, which is an important asset of the healthcare sector.

A blockchain is a type of immutable ledger that is distributed and replicates transactions across its network, incorporating cryptographic links in chronological order between information. It consists of consensus protocol and smart contracts to ensure security. It can overcome the obstacles experienced by AI by establishing trust among users, organizing data, and facilitating resource sharing in AI-based healthcare. Peer-to-peer networks and decentralized ledger is a feature of blockchain technology. Transactional records are securely maintained by a distributed ledger. By logging local gradients on the blockchain, this functionality supports the safe learning of heterogeneous data. Similarly, the execution of transactions in a distributed network without the involvement of a third party or centralized authority is done automatically by a smart contract. It is a piece of executable code that is present on every node and that is triggered when a transaction is initiated. The transaction is validated via smart contracts. Smart contracts enable the imposition of access control regulations for data access. Smart contracts enable user provenance. In the case of transactional data, a block is produced. A consensus algorithm is used by miners to commit the block to the blockchain. Algorithms of consensus mine the block. It forces miners to work through challenging cryptographic riddles and publish their solutions with other miners. The opportunity to mine a block of the transactions and adding it to the available chain and duplicating the created chain in all the connected nodes is given to the miner who solves the challenge first. Consensus algorithms may be the effective method for group decisions on diagnosis and treatment in AI-based healthcare systems. Blocks are immutable and auditable since they are cryptographically connected to one another. When the transaction is copied and duplicated across all network nodes, the highest level of availability and transparency is achieved. Medical data can be verified via cryptographic linking, which can also provide a tamper-proof duplicate of it. Anyone can join the network and take part in transactions while using a public blockchain. Private blockchain, on the other hand, places restrictions on access without sufficient authentication and verification. Public and private blockchain characteristics are combined in consortium blockchain. The working of the blockchain is shown in Fig. 2.3.

Figure 2.4 shows some of the applications of blockchain in healthcare. Securing patient's medical data and effectively managing various product supply chains, including those for medical equipment, organs, medicines, drugs components, oxygen cylinders inclusive of all other pharmaceuticals, are two essential criteria of the healthcare business.

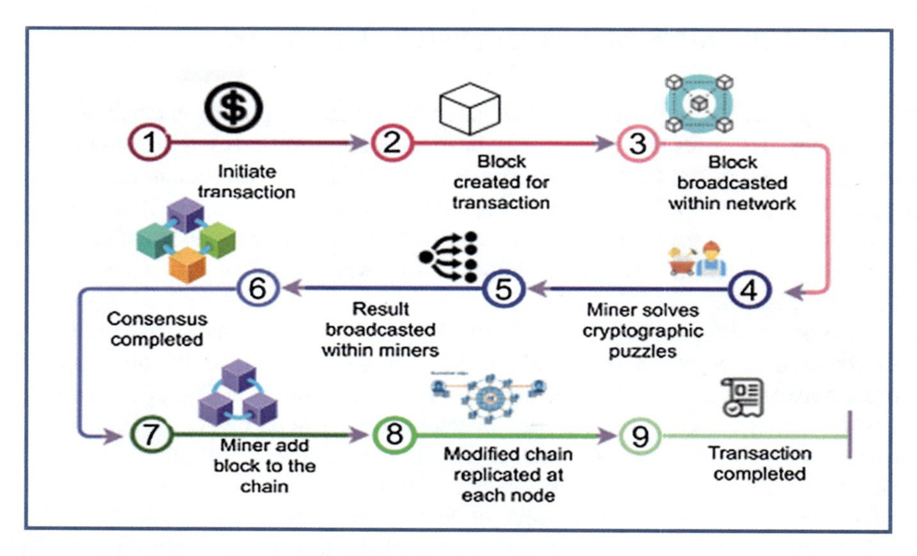

Fig. 2.3 Process workflow involved in blockchain

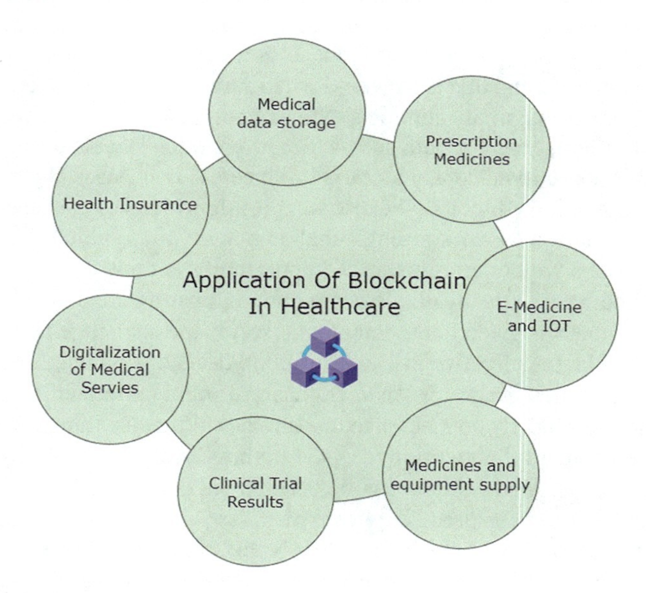

Fig. 2.4 Blockchain and healthcare

Blockchain is a leading-edge technology that has the potential to revolutionize the healthcare system by providing security, dependability, confidentiality, and compatibility [2]. It features an unalterable and distributed ledger in which patients' medical data can be stored securely and prevented from tampering. It is safeguarded by cryptographic elements like hashing, digital signatures, and asymmetric keys ensuring that the data cannot be tampered with [3]. Since the ledger is decentralized, any slight alteration to a data transaction will be detected by all blockchain members, resulting in greater transparency across the entire system. Using blockchain technology enables safe medical data transmission, prevents breaches, and effective management of medical resources as the healthcare sector is constantly at risk of being attacked. A healthcare system that uses blockchain technology was presented by the authors of [4] to protect the confidentiality of user data. They also utilized various mechanisms to safeguard users' confidential information, developed smart contracts in order to authenticate transactions of data, and provide control access and decision-making in an open network. A safe and dependable blockchain-adapted strategy to prevent security violations of electronic medical record systems was put forth by Ray et al. in [5]. To enable secure data sharing over the IoT network, they deployed private blockchain and swarm intelligence techniques. Moreover, Subramanian et al. [6] examined the use of blockchain and AI technology in the treatment of diabetes disorders, particularly during the COVID-19 pandemic. Similarly, medical facilities, testing facilities, academic institutions, and patients may share useful information and collaborate to enhance the AI model. Nevertheless, due to privacy and security issues, they have trouble sharing crucial data with outside parties. Hence, a barrier to raising the caliber of AI-based healthcare systems is secure data sharing. In order to improve the prediction of lung cancer using CT scan pictures, Kumar et al. [7] suggested a method that involves exchanging regional models through the network of blockchain. Hence, the updated model assists in precisely diagnosing the ailment of the patients, leading to enhanced therapy. By preventing actual data sharing, privacy is maintained. Organizations will exchange local gradients via smart contracts and transfer their data to the IPFS (Interplanetary File System). The global model is trained using a consensus approach called Delegated Proof-of-Stake. A smart contract establishes trust in the data, and the blockchain is updated with the local gradient's hash. To expedite biomedical research, Mamoshina et al. [8] have offered AI and blockchain technologies. It also provides patient incentives to get regular examinations and benefits from new technology for managing and making money through their personal information. Patients can sell their medical records using tokens on the permissioned blockchain platform called Exonum, which has been proposed by the group. Nevertheless, once data has been sold to authorities, this framework has no control over it. An intrusion detection system has been proposed by Nguyen et al. [9] to safeguard data transfer in the healthcare industry's cyber-physical system. Patients frequently lack control over who has access to their medical data. A safe, immutable, and decentralized gradient mining is used in place of the insecure central gradient aggregator on the blockchain. Smart contracts are used to control the edge computing, management of trust, authentication, and distribution of trained models, as well as the identification of nodes and the datasets or models used by them. This method offers total encryption for both a trained model and a dataset. A decentralized AI-powered healthcare system has been built by Puri et al. [10] that can access and verify IoT devices along with fostering trust and transparency in the health records of patients. This method uses the creation of a public blockchain network and AI-enabled smart contracts. The framework also finds IoT nodes in the network that might be harmful. BITS is a special intelligent TS system built on blockchain that is offered by Gupta et al. [11]. They offer extensive insights into the blockchain- and cloud-based smart T's frameworks, highlighting the difficulties with data management, security, dependability, and secrecy.

To maintain the security and privacy of the IoMT, Polap et al. [12] have provided distributed learning. It utilizes decentralized learning along with blockchain security enabling the creation of intelligent systems that preserves confidentiality by keeping the data locally stored. The model poisoning attacks can be lessened by using this approach. Kumar et al. [13] described a method to detect people infected with COVID-19 through CT images by developing a model jointly using blockchain technology and federated learning to maintain secrecy. To solve these challenges, Chained Distributed Machine Learning (C-DistriM), which is a unique decentralized learning that also uses blockchain-based architecture, has been predicted by Zerka et al. [14] to be made for imaging in the medical field. Blockchain preserves model integrity and records the unchangeable history of computation. The Explorer Chain framework, which was proposed by Kuo et al. [15], aims to build a model that can predict throughout the distributed architecture. The framework employs machine learning and blockchain technology that does not require patient data sharing or a central coordinating node, making it decentralized and without a central authority. Similarly, in order to establish the transmission of data, transfer of the model, and its testing in three places in China and Singapore, Schmetterer et al. [16] implemented a blockchain-enabled AI technology. A wireless capsule endoscopy approach for identifying stomach infections was investigated by Khan et al. [17]. A complex artificial neural network model is secured using a blockchain-based method to enable accurate diagnosis of gastrointestinal conditions like tumors and bleeding. Each part includes a separate block that stores specific data to fend off attempts that would temper or modify it. Natural language processing (NLP) technology, in particular, has proven an efficient tool to categorize the emotion, and feelings of texts, present in social media such as posts, according to Pilozzi et al. [18]. These methods could be applied to learn more about how people see Alzheimer's disease. Patients will have more control over their data if decentralized, secure data transit and storage techniques like blockchain are used. Most of the anxieties associated with mistakenly revealing personal information to an organization that might treat the patient unfairly will be eliminated. The work that has been done in blockchain for AI-based healthcare is shown in Fig. 2.5. It shows the kind of blockchain that is used for the various data modalities. Blockchain is categorized into three types: consortium, private, and public blockchain. The most used public blockchain is Ethereum and the private is Hyperledger.

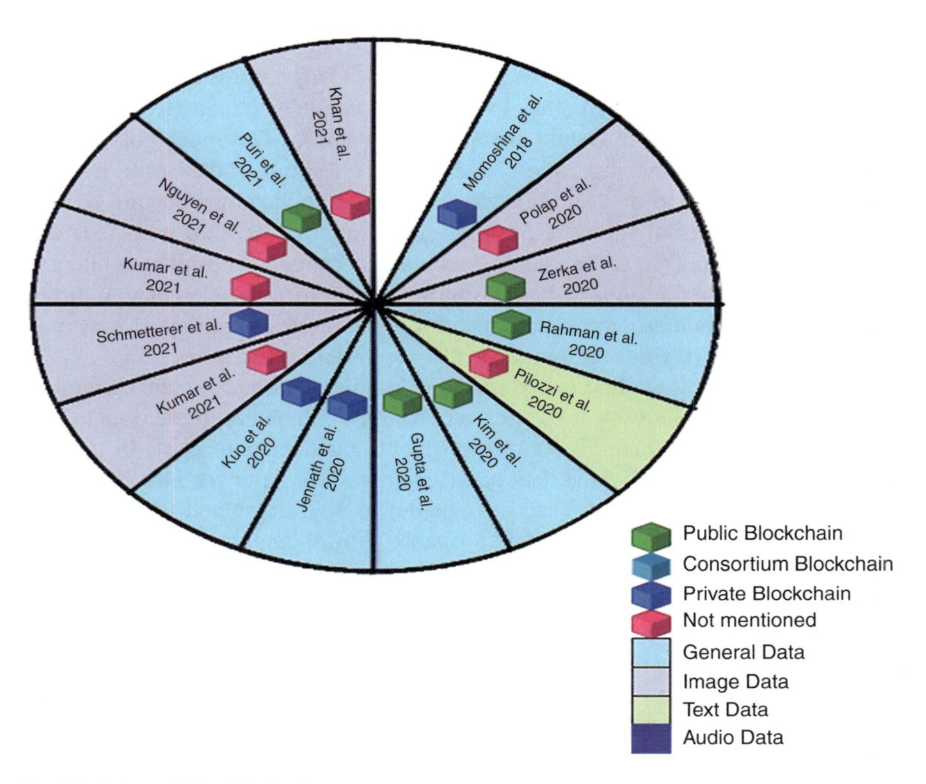

Fig. 2.5 Types of blockchain, data

4 Role of Artificial Intelligence in Smart Healthcare Systems

The healthcare sector demands advanced and anticipatory solutions that offer boundless prospects for accurate and beneficial patient treatment and management operations [19]. IoT technologies generate a vague volume of data and transfer it to and from different parts of the health industry. The health industry needs to implement AI technologies for the effective management of data and its improvisation. The employment of different AI devices in the health sector has many benefits over the current system, which relies on time-consuming data analysis and decision-making methods. In order to offer insightful information about diagnosis, clinical decision support, and treatment, it interacts with medical data.

For instance, in [20], scientists looked at the osteoporosis condition, which is typically identified by conventional X-rays and MRI scans. An AI-featured selection technique was used by the authors to facilitate the diagnosis of osteoporosis patients through the data obtained by ultrasound. As a result, they were able to classify osteoporosis patients' fracture risk with 71% accuracy. Wazid et al. [21] discussed the key features of AI technologies in the healthcare industry. They employed AI algorithms to effectively forecast the likelihood of myocardial infarction and the possibility of developing tumors while also uncovering insightful patterns in the medical data. In [22], Parra et al. investigated how AI algorithms could be used for sustainable development. Here, they looked at the people who required an AI-based question-recommendation system for various scenarios. The main goal of their study was to support the suggestion for AI-based questions in the health industry, having the potential to be used in a vast number of applications beyond security screening and financial services. Similarly, Tedeschini et al. [23] used federated learning, a decentralized technique, to create a distributed networking architecture for segmenting brain tumors based on message queuing telemetry transport (MQTT). Their findings demonstrate that the suggested framework performs more accurately and quickly during routine healthcare system activities.

A data processing method called machine learning automatizes the creation of models capable of analysis. It focuses on enabling computers to study data, find patterns, and make human-like choices without actual human input. In order to complete these tasks, validated data had to be obtained. After the classifiers were successfully trained, the model had to be deployed. Retraining and feedback loops may be used to continue improving performance. Any attempt to uncover, alter, disable, harm, capture, or collect information by taking advantage of system weaknesses constitutes a threat to the device. The fundamental security requirement for any system is to maintain the privacy of sensitive data or processes. In order to keep the trained model from malfunctioning, three vitals must also be safeguarded.

The section that comes after Fig. 2.6 focuses on the AI attack surface. Any adversary can attack an AI-based system by targeting data, classifiers/algorithms, and learning models.

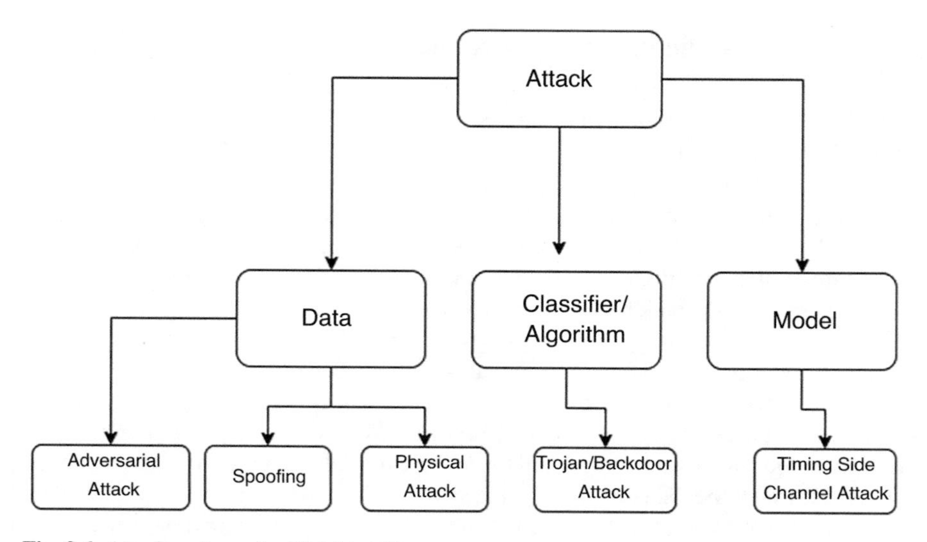

Fig. 2.6 Attack surface of artificial intelligence

4.1 Data

Data is the raw statistics and facts used by a machine. It is a critical component of artificial intelligence. Data is required to train current models and development of all current technologies. It costs a lot of money just to get as much precise data as you can. The attack's data-targeting strategy has a significant impact on AI-based systems. By taking advantage of the extraordinary sensitivity of AI to detect slight differences in the input, known as a poisoning attack, data can be violated either during the learning phase or during the filed-test. These assaults may be promoted via spoofing [24]. It is a type of cyberattack when a malicious party uses a computer, device, or network to pretend to be someone else in order to trick other computer networks. Malicious opponents typically cannot access the training phase of the model. In order to trick a classifier or avoid being detected by a neural network during testing, they produce hostile input. These attacks can be of the physical or digital variety. For this study, we are concentrating on cyberattacks of various kinds. A digital technique immediately introduces small input perturbations. In this case, the attacker can take advantage of the system that has been targeted without the detection system noticing. Concept drift might also result from evasion attacks [25]. Prospective attackers may potentially acquire access to the training datasets and conduct poisoning attacks, which contaminate the datasets with adversarial samples. As will be covered in more detail in subsequent parts, adversarial attacks can cause potential damage to the system.

4.2 Classifier/Algorithm

Classifiers/Algorithms are usually affected by a Back-access Attack. A Trojan assault undermines the authentic model by incorporating a secret entrance to the neural network, which is triggered by a specific pattern in the testing data. This will alter the network using a compromised dataset [26]. Trojan assaults vary from adversarial attacks even though both only occur during the training phase. An adversarial attack merely influences the outcome in this scenario rather than forcing the neural network to change itself. But, a trojan assault, causes the network to change itself because of the poisoned input samples so that it can accurately function for benign input samples. As a result, the network will only malfunction when a trojan causes it to. A user may have trouble recognizing the trojan assault [27]. A trojan attack may result from a SPA (Stealthy Poisoning Attack), which is dependent on a Generalized Adversary Network (GAN) [28]. Another illustration of a neural trojan assault is Badnet [29]. The situation of a trojan assault is shown in Fig. 2.7.

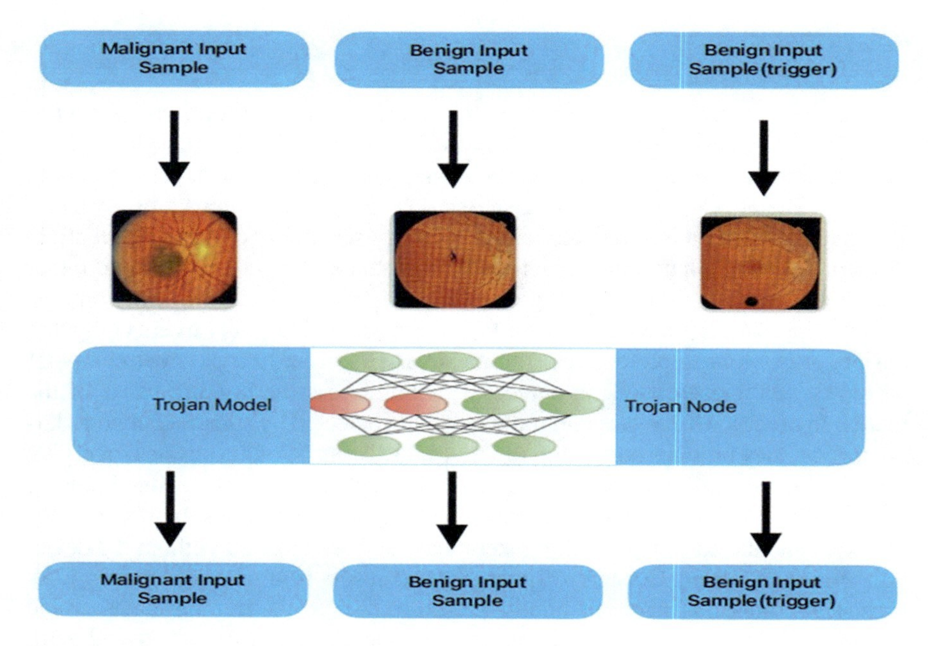

Fig. 2.7 A sample example of a trojan attack

4.3 *Model*

The model is affected by timing side-channel attack. Neural networks are susceptible to attacks like timing side channels because of their peculiar qualities of having varied execution times depending on the depth of the network. An opponent can determine the layers of the neural network, by observing how long the model takes to generate output. This adversary makes use of a regressor that was trained using various network layer counts and execution timings. The information is then used to create replica models that have features in common with the original network [30]. The essential elements are retained due to memory access patterns. Reverse engineering of the CNN model's structure and weights can be used to leak information through memory and timing side-channel attacks. The crucial characteristics of a neural network [31], such as the overall layers, the size of each layer, and the interdependencies among them, are exploded by the memory access patterns.

5 Machine Intelligence Technologies in Healthcare

5.1 *Text-Based AI Technology*

The text-based AI technology widely used is natural language processing. It is a machine learning model, which aims to enable computers to understand, process, and generate human-like text and language. The goal of NLP is to build systems and algorithms that can perform a variety of tasks involving human languages, such as text classification, machine translation, sentiment analysis, recognition of speech, and natural language generation.

5.1.1 Applications in Healthcare

Clinical Decision Support Systems (CDDS) receive a variety of inputs, including incomplete structured data like XML files, structured data like HER, and unstructured data like diagnostic summaries and progress records. To aid clinical decisions, various systems have been introduced that utilize NLP techniques that take input from unstructured data, specifically for the purpose of calculating and automating diagnoses or treatments. With the help of NLP, CDSS can create results and recommendations that help healthcare practitioners make the best decisions possible by automatically extracting key information from free text [32, 33]. NLP makes it easy to extract important clinical information from unstructured data in medical records, such as physician notes, discharge summaries, and diagnostic reports. This can help with coding, billing, and clinical decision-making. A sentiment score system has been used to assess sentiment statements of admission and discharge in a hospital [34]. Unstructured reports are also used for radiology. NLP enables the recognition of key aspects in those reports, their extraction, and conversion into usable computer formats [35]. The analysis of vast amounts of free-text medical reports using NLP contains the potential to contribute to the development of procedure-intensive fields such as Hepatology. Also, NLP can be used to develop chatbots and virtual assistants that can answer patient questions and provide basic medical advice. This can help patients access healthcare information quickly and easily [36]. In a recent study, NLP was used to categorize diseases and conditions which were challenging to identify through simple clinical procedures. Using NLP-based solutions for information retrieval (IR) reduces the time and effort required, ultimately promoting the therapy [37, 38].

5.1.2 Adversarial and Defense Attack

An alteration in the text's semantics, grammar, or visual similarity that deceives NLP is known as adversarial text. The techniques for creating hostile text are shown in Fig. 2.8. To impact the model's prediction, text-based adversarial examples can

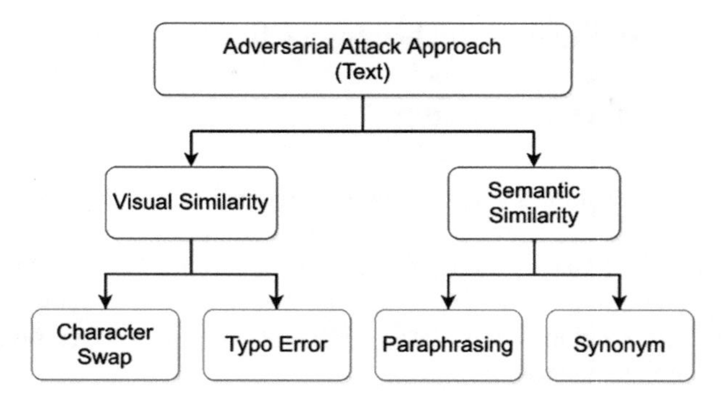

Fig. 2.8 Taxonomy of adversarial attack on text

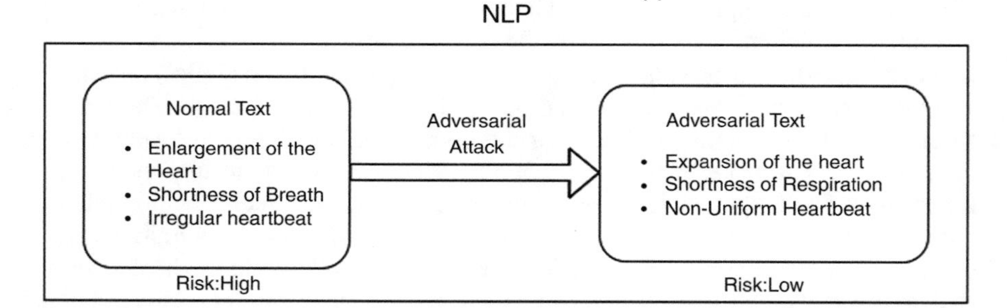

Fig. 2.9 Adversarial attack on NLP-based healthcare application

be created by making slight changes to the text, mimicking real typing errors made by humans. These modifications are intended to be minimal yet effective in altering the model's output. After the attack, the text appears fairly near to the original. This method of attack is used in assaults like hot flip [38], textbugger [39], and DeepWordBug [40]. The process of making hostile writing involves paraphrasing the source material. The semantic equivalent of the original text will be produced by this assault, but the model's result for the original text and the modified text will be different.

Figure 2.9 shows how NLP-based healthcare is affected by adversarial attacks. The NLP can be tricked by simply substituting synonyms for words while keeping the text's semantics. The incorrect diagnosis ultimately results in the incorrect treatment, endangering lives.

5.1.3 Blockchain-Based Solution for NLP-Based Healthcare

As mentioned above, NLP faces different types of adversarial attacks. Contrarily, those attacks are only somewhat sophisticated; in the case of text data, these attacks are plainly visible to the naked eye. So, we might draw the conclusion that there is a moderate chance that such attacks will occur. In order to address attack surfaces including data, classifiers, and models in NLP, we have framed blockchain solutions.

Data Layer

In NLP, data may be stored locally on the computers of data owners, such as physicians, hospital staff, and laboratories database. A peer-to-peer blockchain network can be created with dispersed owners to look into the issue of having enough data for training AI models while also maintaining the secrecy of the data by enabling owners to transfer their data indirectly with other parties. Off-chain data storage is supported by this framework. This peer-to-peer network provides direct service exchange using a suitable authentication method. Without a centralized server, thousands of devices can be linked together. The P2P blockchain node can take on the role of a service provider or requester. Rules can be inferred for access control to allow private data sharing through smart contracts. To verify the integrity of distributed data, a hash code is generated, which is recorded in the blockchain at every data center. When data is used for training, the hash will be regenerated and verified using blockchain technology. Figure 2.10, which follows, provides an illustration of datasets construction of NLP using blockchain.

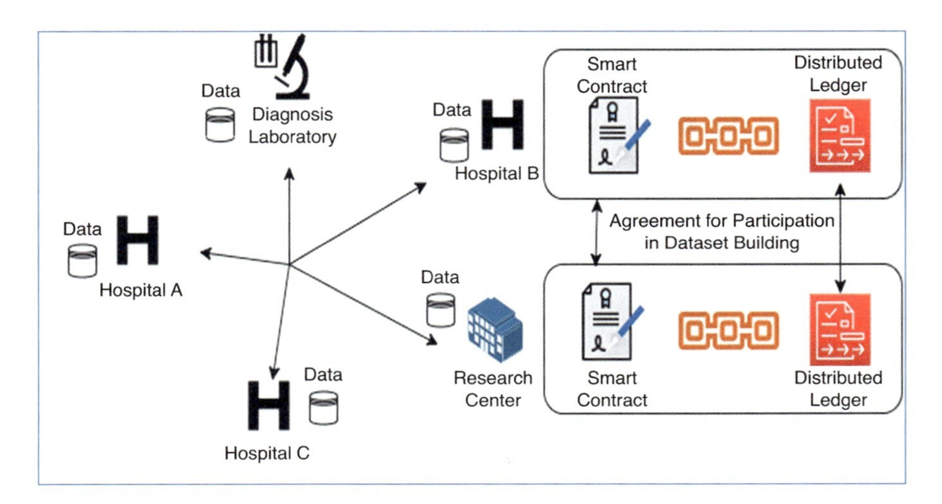

Fig. 2.10 Datasets for NLP-based healthcare using blockchain

For instance, if the research center requires access to the hospital's datasets, it will first submit a smart contract request for data access. Hospitals can reply by stating their permission to take part in dataset creation and providing any guidelines or limitations for data sharing and AI model training. A copy of the distributed ledger is accessible to each participant. In order to do future integrity checks on the data at each station, a hash code can be created, which can be saved in the blockchain. With the help of blockchain, a large number of data stations can contribute to the creation of verified datasets.

Learning Phase

In order to develop an ML algorithm via the available data, distributed learning is used. Threats to federated learning include a broken node, trusting local gradients, and aggregating gradients globally. In federated learning, utilizing blockchain helps address these issues and defend the model from poisoning assaults. A smart contract can be used to start training so that it can verify the legitimacy of the participant. Then, through the block, local gradients at federated nodes will be transmitted. Local gradients will be secured on the blockchain to prevent modification and used for verification later. Using a consensus algorithm, blockchain network miners will validate and produce global gradients. This is how blockchain may give the federated network validity. Each node saves the retrieved features in the distributed ledger for later use and embeds them in vector space. The blockchain approach for the security of the classifier in NLP is shown in Fig. 2.11.

The trained model's output is influenced by how real the post-training input is. We could anticipate the NLP model malfunctioning for adversarial text input. We can attempt to reduce some adversarial assaults by utilizing the blockchain in NLP-based healthcare. When identifying hostile text, the version of characteristics collected from a dataset of blockchain is utilized for model training. A smart contract will produce word embedding for the supplied input. It will search the distributed ledger for a similar corpus of word embedding based on synonyms. The blockchain network's miners then receive additional distributions of the resulting extracts. Miners will use a trained model to compute the outcome for assigned characteristics rather than using proof of work. After that, the result is distributed in the mine pool, and if the majority of them agree, then the outcome is consensually added to the chain. Hence, the model will be protected using this framework from adversarial assaults on text. Figure 2.12 represents the mentioned framework.

5.2 Machine Learning for Medical Imaging

Machine learning applications such as computer vision have been used in healthcare for medical imaging. Computer vision is a visual application of AI and computer science that aims on enabling machines to decipher, understand, and analyze

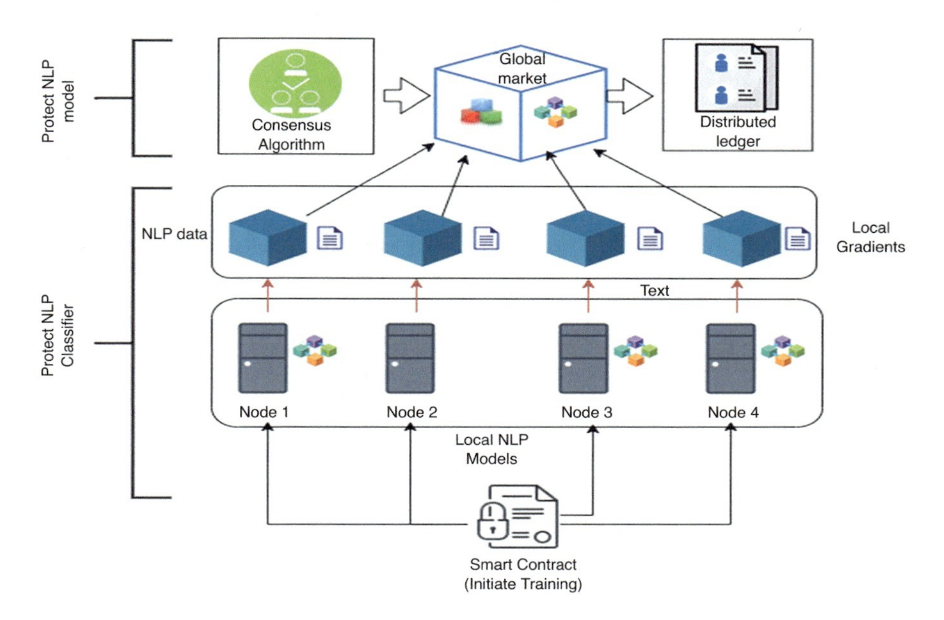

Fig. 2.11 Protection of training phase of NLP using blockchain

perceptible data from the world [41]. It involves developing algorithms and techniques to enable computers to recognize and classify objects, understand scenes, track motion, and more using images and videos. The techniques involved in computer vision are object detection, image classification, object tracking, and semantic and instance segmentation.

5.2.1 Applications in Healthcare

The interpretation and analysis of many types of real-world data are aided by intelligent intervention employing a brain-like structure and advanced technologies like machine learning and computer vision [42]. A scientific application of machine learning is computer vision, which uses collected sequences of movies and photos to identify things. Convolutional neural networks (CNNs), a machine learning algorithm created to analyze picture input, prioritizes different elements to identify one image from another. Similar to the connection pattern of neurons in the brain, CNNs have a structural design. Computers have long been able to analyze visual imagery in meaningful ways thanks to computer vision. Object classification, localization, and detection are the terms used to describe the processes of determining an object's kind, location within an image, and both concurrently [43].

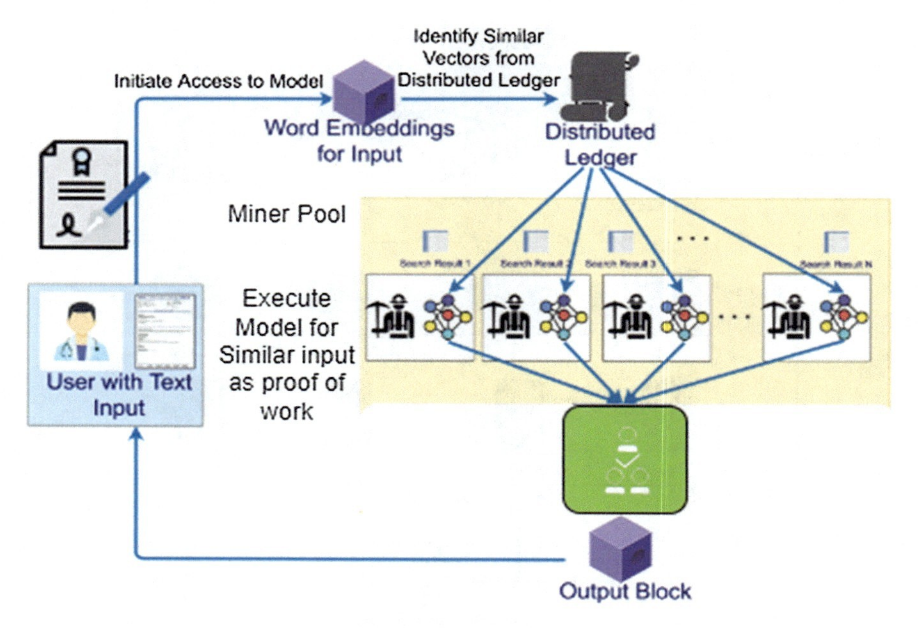

Fig. 2.12 Solution for NLP-based healthcare using blockchain technology

According to the National Cancer Institute's National Lung Screening Trial (NLST), low-dose CT, which is used for screening lung cancer, has caused a 20% reduction in mortality [44]. The use of smart monitoring has increased because of developments in computer vision. To anticipate generalized anxiety disorder (GAD), a new system using computer vision and ML is introduced [45]. Computer vision algorithms in an adult intensive care unit can recognize patient movement actions like getting the patient in and out of bed or a chair [46]. A substantial possibility exists for deep convolutional neural networks as a tool for ear-related diagnosis [47]. A comprehensive image processing system to forecast the viability of human embryos, researchers coupled computer vision methodologies with machine learning and different techniques involving neural networks [48]. Using a computer vision technique, it is also possible to identify hip fractures from pelvic X-rays [49]. Recorded endoscopic pictures will be swiftly and precisely analyzed by the groundbreaking CNN approach to detect esophageal cancer [50]. Deep learning methods also enable the detection of intracranial hemorrhage (ICH) [51]. It is possible to make a diagnosis based on chest CT pictures, leading to a machine learning algorithm in a quick and automated diagnostic technique [52]. In order to decrease the chances of infection from the doctor to the patient COVID-19, a revolutionary visual SLAM algorithm may also follow and find robots in real-time environments [53].

5.2.2 Adversarial and Defense Attacks

The images in which pixels are purposefully disturbed to confuse and deceive models while appearing correct to human sight are adversarial images. Adversarial images trick DNN because it is vulnerable to even the smallest input disturbance. Figure 2.13 displays the ways adversarial attacks can be done on an image.

Several attack strategies, such as FGSM [54], BIM [55], and R + FGSM [56], cause the ML model to make wrong predictions and decreases the overall robustness of the model. Figure 2.14 provides an example of an adversarial attack on an X-ray image.

5.2.3 Blockchain Solutions for Computer Vision–Based Healthcare

Data Layer

Since adversarial attacks are more likely to target images, we frame solutions that emphasize preventive actions, as seen in Fig. 2.15. A blockchain-based system will be used to post images to IPFS. A file-sharing technique called IPFS can be used to store and transport large data. It uses cryptographic hashes, which can be stored in the blockchain easily. The generated hashes are utilized to ensure the authenticity of images. First, different hospitals and diagnostic centers that have the data will request to upload medical images on IPFS, which will be validated by a smart contract before uploading. Every image will have its unique hash, which will be stored in the blockchain. Users can approach IPFS with the hash code to access the image data set when needed. The detection of adversarial images will be easy as hash codes are extremely sensitive. In this way, the data set is secured at IPFS using blockchain.

Learning Phase

Computer vision uses dynamic data for visual inputs. The model is trained with image data sets. As shown in Fig. 2.16, security can be provided through blockchain. With proper authentication, the training in the research centers should be

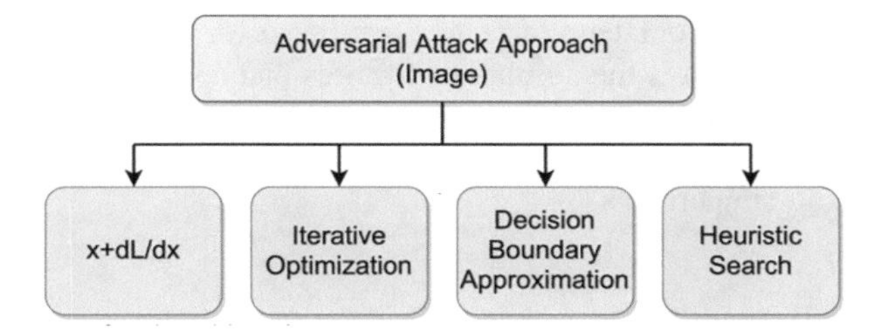

Fig. 2.13 Adversarial attack on image

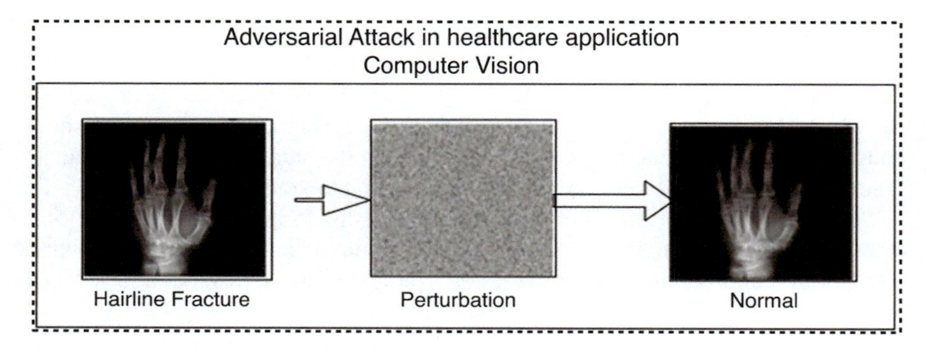

Fig. 2.14 Adversarial attack on computer vision–based healthcare application

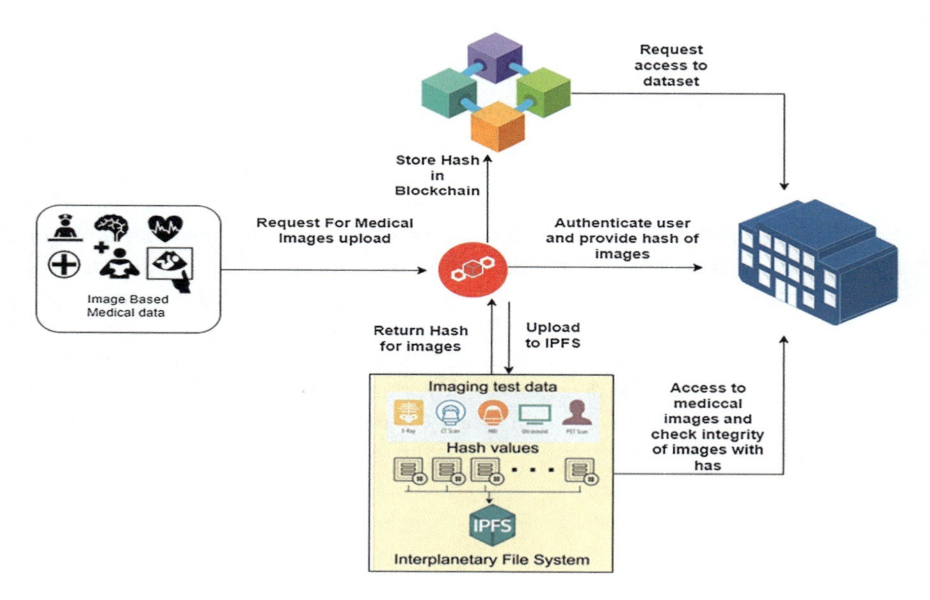

Fig. 2.15 Creating datasets for computer-based healthcare applications

started using smart contracts. After the training, the features that are extracted are stored in the blockchain for later referral as a feature vector with the formula x = (x1, x2, x3,..., xn) T, where n is the number of features that are extracted, and T is the transposition operation. This architecture safeguards the whole computer vision training area. As a result, the features that the learning process retrieved will be preserved without tampering.

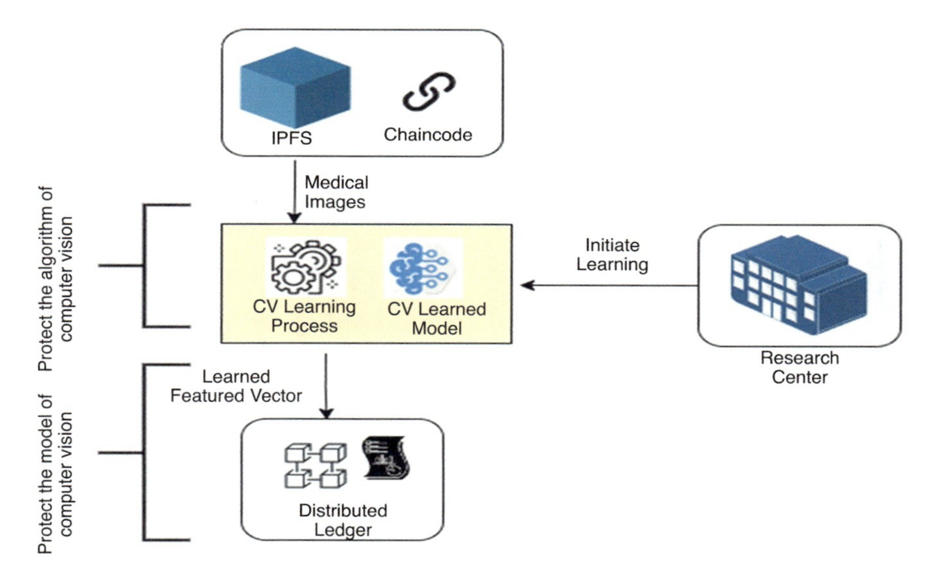

Fig. 2.16 Training phase of computer vision and its protection through blockchain

The outcome generated should be clear and understandable, and it should provide reasons or evidence to support the conclusion. The ability to use the model post-training is limited and controlled using smart contracts. With smart contracts, access to the trained model is controlled. Only licensed physicians and researchers have access to the model. To verify the precision of the model run, it will compare it to the feature vectors recorded in the distributed ledger. Providing limited access to the data will help to tamper-proof it and also make it available when needed. The metadata can be stored in the blockchain for further validation and verification. Fig. 2.17 shows the security of trained models through blockchain.

5.3 Audio-Based AI Technology

Acoustic AI techniques are sound recognition AI technology that uses sound data to identify and classify sounds. These techniques have a huge potential in the healthcare sector, such as in diagnostics, monitoring, and treatment. Acoustic AI techniques have become widely used in diagnosis and treatment in healthcare. Some of the techniques involved in acoustic AI are selective noise canceling, Hi-fi audio reconstruction, analog audio emulation, speech processing, and improved spatial simulation.

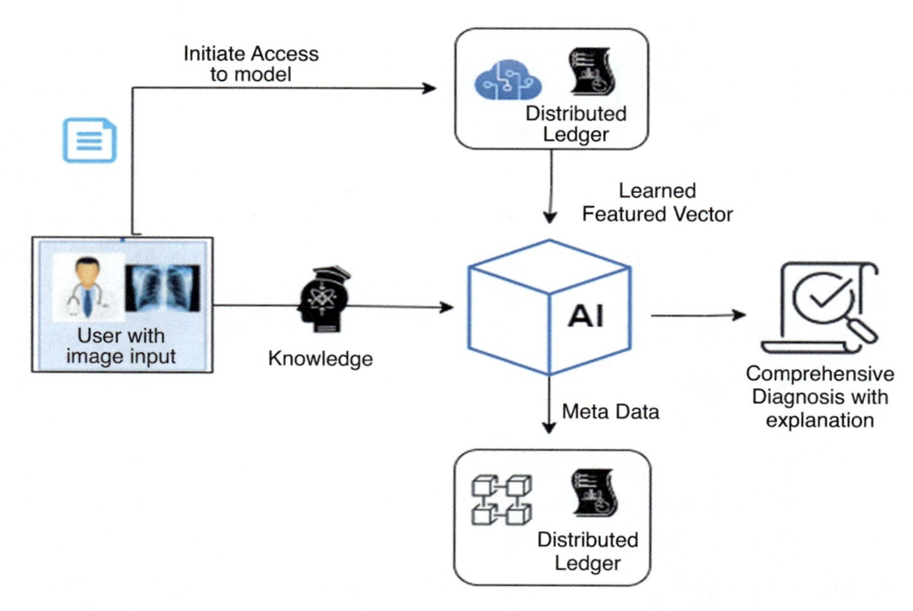

Fig. 2.17 Blockchain security for a trained model of computer vision

5.3.1 Applications in Healthcare

Numerous software applications employ sophisticated AI algorithms and process digital signals to identify complete sleep schedules, track the rate of breathing, detect gasping and snorting, and recognize patterns of sleep apnea. These applications then utilize this information to accurately measure a person's respiratory rate while they are sleeping, all via smartphones. These applications combine active sonar and passive acoustic analysis. One of the application frameworks is "Firefly" [57]. A framework designed using neural networks (NNs) can distinguish between four different forms of auscultatory noises, including wheezes, rhonchi, fine crackles, and coarse crackles, which reduces human mistakes during auscultation [58]. Researchers have developed classifiers using this technology that can distinguish between different respiratory illnesses in adults using the auditory features of coughs. Also, they have created synthetic cough samples for each significant respiratory ailment, using recent advancements. To help doctors, machine learning algorithms identify the earliest stages of pulmonary disease, for example, Cough GAN generates simulated coughs that mimic major pulmonary symptoms. By accurately and early diagnosing advanced respiratory illnesses such as chronic obstructive pulmonary disease, doctors will create the best preventative treatment programs and lower morbidity [59]. AI-based technologies are used for pediatric breath sound classification where the use of a CNN architecture (N-CNN) along with other CNN architectures can be applied to examine discomfort in babies through their sound of crying patterns. Results show that this method is a much more beneficial and viable, alternate to the method of evaluation used conventionally [60].

5.3.2 Adversarial Attack

Adversarial audio is any audio that contains disturbance, or noise, often known as adverse perturbations, and it can fool a variety of sound classification systems. Figure 2.18 shows the classification of adversarial attacks in audio signals. Some assaults aim to create an adversarial audio sample that closely resembles the original, but the learned model would classify it incorrectly. These assaults fall under the category of speech-to-label assaults. By using the actual audio and the required output label, the attacker can show how genetic algorithms can generate hostile audio samples without the use of gradients. It increases random noise while preventing human awareness of it [61]. During the conversion of speech to text through acoustic processing, an adversary can attempt to manipulate the output to achieve a specific result. Such attacks are referred to as speech-to-text attacks. It is possible to alter the audio spectrum to obtain a desired output by introducing a minor disturbance using optimization-based attacks [62]. Figure 2.19 provides an example of an adversarial attack on an acoustic technology application in the healthcare system.

5.3.3 Blockchain Solutions for Acoustic AI-Based Healthcare

Data Layer

Data resides locally on the computers of data owners, such as physicians, hospital staff, and laboratories database, similar to NLP-based healthcare. Figure 2.20 demonstrates the data layer construction using blockchain for acoustic AI-based healthcare. As there are many IoMT technologies that can threaten the privacy of entities

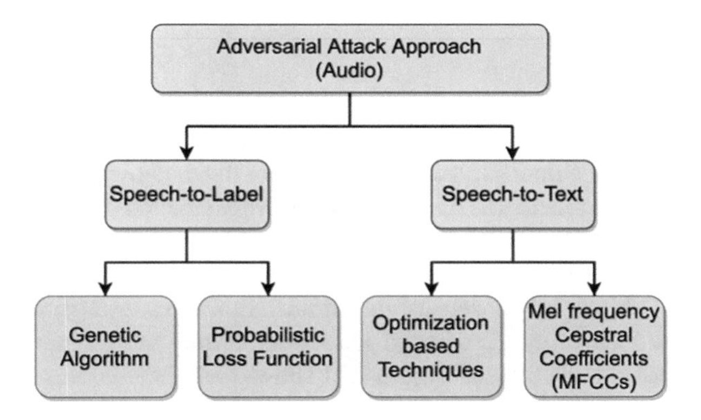

Fig. 2.18 Classification of adversarial attack on audio

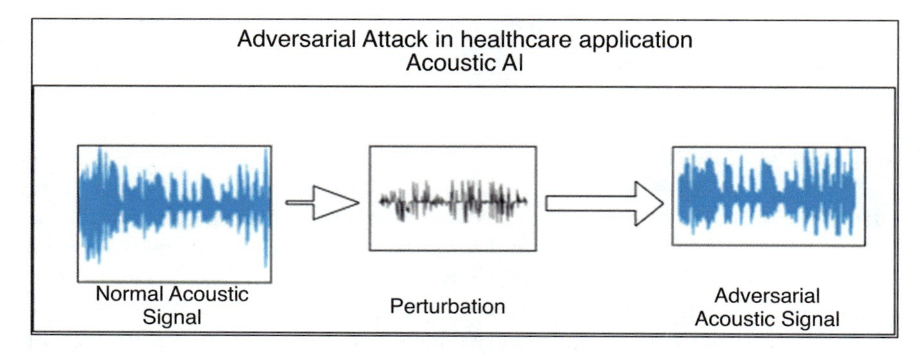

Fig. 2.19 Adversarial attack on acoustic-based healthcare application

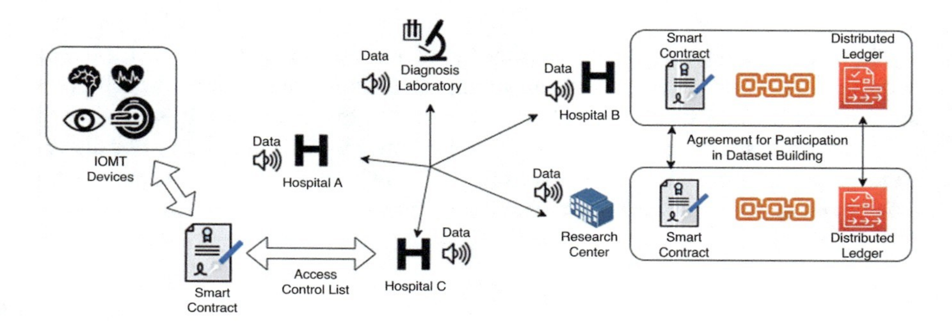

Fig. 2.20 Blockchain and dataset construction in acoustic AI-based healthcare

with data, blockchain can be used to prevent the risks. Smart contracts can be deployed for access control on acoustic data storage. Similarly, IoMT devices can be checked for proper registration and authentication procedure before contributing to the data layer. Sharing of data for data layer construction is similar to the NLP-based healthcare data layer construction framework described earlier.

Learning Phase

In order to train our acoustic AI model, a federal learning approach is adopted as the data is distributed. The nodes in the AI network are the data owners. For subsequent verification and reference, the features that were extracted from audio samples would be safely stored in distributed ledgers. Depending on the learning strategy, the audio sample's extracted features can take on any shape. The job of creating the global model is driven by consensus algorithms, and each local gradient is kept in the blockchain. Hence, the distributed ledger containing the global model can be protected against several threats. Figure 2.21 shows how the model is protected with blockchain technology.

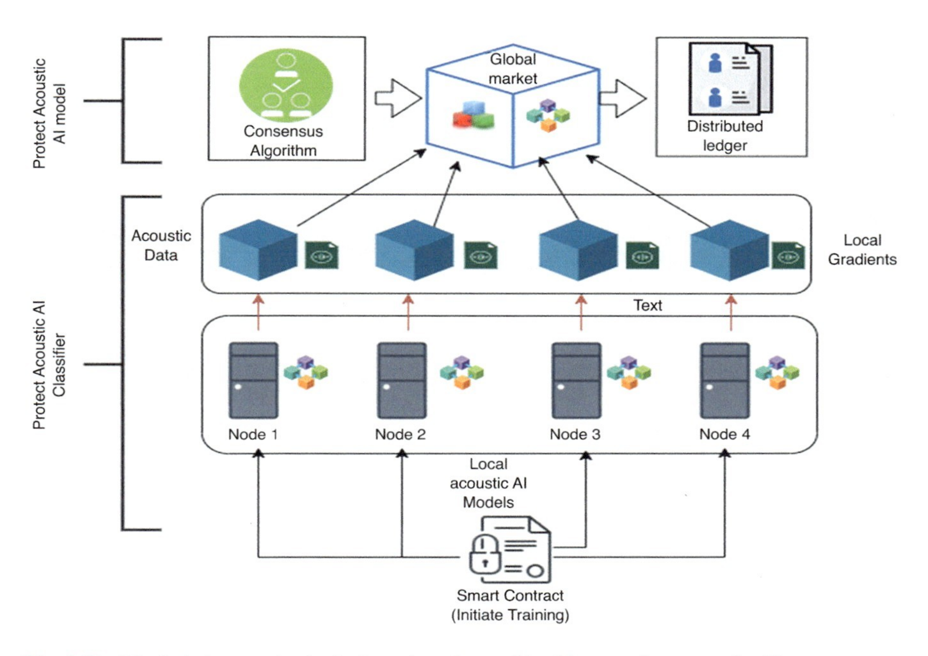

Fig. 2.21 Blockchain security in the learning phase of healthcare using acoustic AI

After training, smart contracts are used to limit users of the model via authentication protocols. A consensus algorithm is used to check if the input is legit or has been tampered with as audio signals are static dependent, that is, the previous behavior impacts the current behavior. Figure 2.22 represents the blockchain framework designed to protect the model of acoustic AI after the completion of the learning phase. The user can get access to the model through a smart contract. It then proceeds to the consensus algorithm of the blockchain network where it is broken into N numbers and each fragment is given to the miners. The result is combined after the mining is completed. If the data has been adversary, then the result will not make any sense as it has a static dependency. Hence, it will help to detect any adversarial attack on the model.

6 Conclusion

This chapter provides an outlook on AI-based healthcare technologies and their security through blockchain. Several research has been conducted in the field of AI and blockchain and their application. In this review, we have discussed different

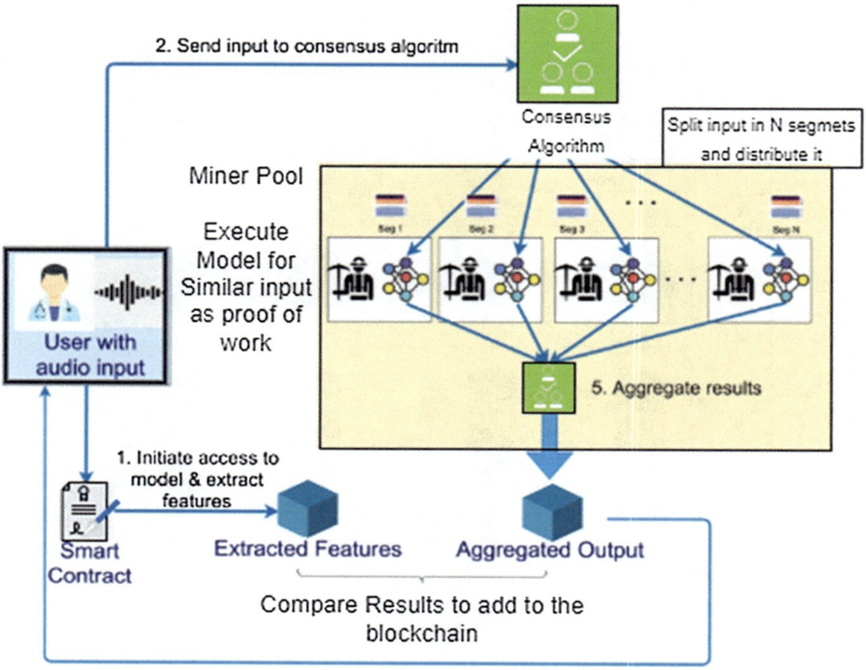

Fig. 2.22 Acoustic AI in healthcare and its security through blockchain

fields of AI, which include machine learning through textual data (natural language processing), medical imaging, and acoustic AI in healthcare. We have also discussed in our review various adversarial attacks and threats these sectors might face and its solution using blockchain technology. The potential that blockchain has in regard to security is undeniable. Blockchain is an immutable, distributed, decentralized ledger that contains the vast potential to safeguard the health sectors against different kinds of adversarial attacks, and privacy issues they might face during data storing and sharing. Figure 2.23 shows the overall properties and application of blockchain. In this chapter, we have discussed how blockchain can provide security in the data layer and training phases of the field mentioned. We have referred to various articles and review documents to collect information and conduct this research. Future research directions have been presented in this chapter for using blockchain in the field of AI and healthcare, which was developed through knowledge and information from current technologies, their application, threats, and existing challenges.

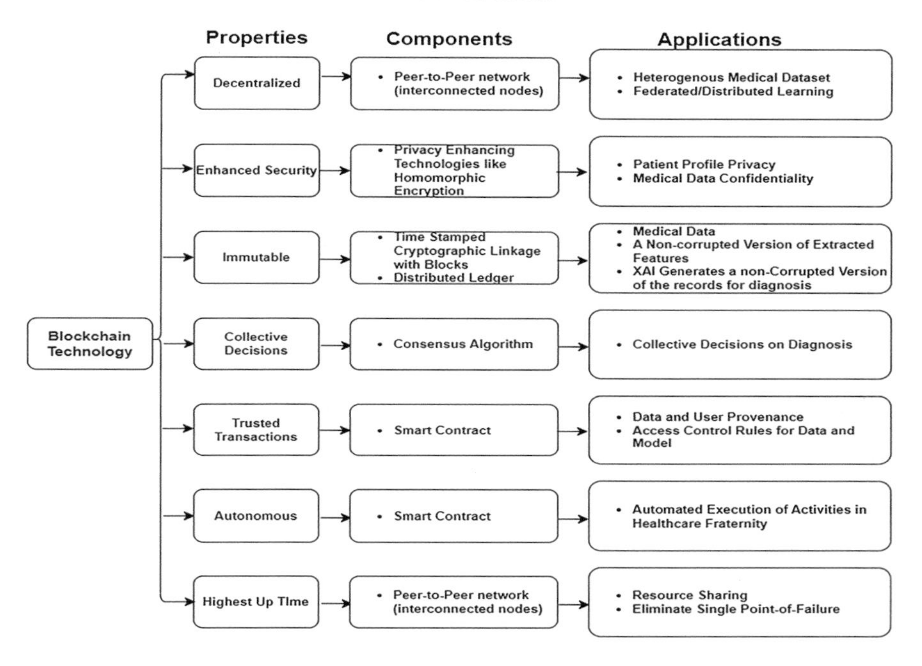

Fig. 2.23 Blockchain for AI-based healthcare explained

References

1. Yaeger, K., Martini, M., Rasouli, J., & Costa, A. (2019). Emerging blockchain technology solutions for modern healthcare infrastructure. *Journal of Scientific Innovation in Medicine, 2*, 1–7. [CrossRef].
2. Gupta, R., Reebadiya, D., Tanwar, S., Kumar, N., & Guizani, M. (2021). When blockchain meets edge intelligence: Trusted and security solutions for consumers. *IEEE Network, 35*, 272–278. https://doi.org/10.1109/MNET.001.2000735. [CrossRef].
3. Kumari, A., Gupta, R., Tanwar, S., Tyagi, S., & Kumar, N. (2020). When Blockchain meets smart grid: Secure energy trading in demand response management. *IEEE Network, 34*, 299–305. https://doi.org/10.1109/MNET.001.1900660. [CrossRef].
4. Wu, G., Wang, S., Ning, Z., & Zhu, B. (2022). Privacy-preserved electronic medical record exchanging and sharing: A blockchain-based smart healthcare system. *IEEE Journal of Biomedical and Health Informatics, 26*, 1917–1927. https://doi.org/10.1109/JBHI.2021.3123643. [CrossRef] [PubMed].
5. Rather, I. H., & Idrees, S. M. (2021). Blockchain technology and its applications in the healthcare sector. In *Blockchain for healthcare systems* (pp. 17–25). CRC Press.
6. Subramanian, G., & Sreekantan Thampy, A. (2021). Implementation of Blockchain consortium to prioritize diabetes patients' healthcare in pandemic situations. *IEEE Access, 9*, 162459–162475. https://doi.org/10.1109/ACCESS.2021.3132302. [CrossRef].

7. Kumar, R., et al. (2021, January). An integration of blockchain and AI for secure data sharing and detection of CT images for the hospitals. *Computerized Medical Imaging and Graphics, 87*. https://doi.org/10.1016/j.compmedimag.2020.101812
8. Mamoshina, P., et al. (2018). Converging blockchain and next-generation artificial intelligence technologies to decentralize and accelerate biomedical research and healthcare. *Oncotarget, 9*(5), 5665–5690. https://doi.org/10.18632/oncotarget.22345
9. Nguyen, G. N., le Viet, N. H., Elhoseny, M., Shankar, K., Gupta, B. B., & El-Latif, A. A. A. (2021). Secure blockchain enabled cyber–physical systems in healthcare using deep belief network with ResNet model. *Journal of Parallel and Distributed Computing, 153*. https://doi.org/10.1016/j.jpdc.2021.03.011
10. Puri, V., Kataria, A., & Sharma, V. (2021). Artificial intelligence-powered decentralized framework for Internet of Things in Healthcare 4.0. *Transactions on Emerging Telecommunications Technologies*. https://doi.org/10.1002/ett.4245
11. Gupta, R., Thakker, U., Tanwar, S., Obaidat, M. S., & Hsiao, K. F. (2020, October). BITS: A blockchain-driven intelligent scheme for telesurgery system. https://doi.org/10.1109/CITS49457.2020.9232662
12. Polap, D., Srivastava, G., Jolfaei, A., & Parizi, R. M. (2020, July). Blockchain technology and neural networks for the internet of medical things. In *IEEE INFOCOM 2020 – IEEE conference on computer communications workshops (INFOCOM WKSHPS)* (pp. 508–513). https://doi.org/10.1109/INFOCOMWKSHPS50562.2020.9162735
13. Kumar, R., et al. (2021). Blockchain-federated-learning and deep learning models for COVID-19 detection using CT imaging. *IEEE Sensors Journal*. https://doi.org/10.1109/JSEN.2021.3076767
14. Zerka, F., et al. (2020). Blockchain for privacy preserving and trustworthy distributed machine learning in multicentric medical imaging (C-DistriM). *IEEE Access, 8*, 183939–183951. https://doi.org/10.1109/ACCESS.2020.3029445
15. Kuo, T. T., Gabriel, R. A., Cidambi, K. R., & Ohno-Machado, L. (2020, May). EXpectation propagation LOgistic REgRession on permissioned blockCHAIN (ExplorerChain): Decentralized online healthcare/genomics predictive model learning. *Journal of the American Medical Informatics Association, 27*(5), 747–756. https://doi.org/10.1093/jamia/ocaa023
16. Schmetterer, L. et al. (2021). *Retinal photograph-based deep learning algorithms for myopia and a blockchain platform to facilitate artificial intelligence medical research: A retrospective multicohort study* [Online]. Available: www.thelancet.com/
17. Khan, M. A., et al. (2021). A blockchain-based framework for stomach abnormalities recognition. *Computers, Materials and Continua, 67*(1). https://doi.org/10.32604/cmc.2021.013217
18. Pilozzi, A., & Huang, X. (2020, March 1). Overcoming Alzheimer's disease stigma by leveraging artificial intelligence and blockchain technologies. *Brain Sciences, 10*(3). https://doi.org/10.3390/brainsci10030183
19. Rehman, M. U., Shafique, A., Ghadi, Y. Y., Boulila, W., Jan, S. U., Gadekallu, T. R., Driss, M., & Ahmad, J. (2022). A novel ChaosBased privacy-preserving deep learning model for cancer diagnosis. *IEEE Transactions on Network Science and Engineering*, 1–17. https://doi.org/10.1109/TNSE.2022.3199235. [CrossRef].
20. Miranda, D., Olivares, R., Munoz, R., & Minonzio, J. G. (2022). Improvement of patient classification using feature selection applied to bidirectional axial transmission. *IEEE Transactions on Ultrasonics, Ferroelectrics, and Frequency Control, 69*, 2663–2671. https://doi.org/10.1109/TUFFC.2022.3195477. [CrossRef] [PubMed].
21. Wazid, M., Singh, J., Das, A. K., Shetty, S., Khan, M. K., & Rodrigues, J. J. P. C. (2022). ASCP-IoMT: AI-enabled lightweight secure communication protocol for internet of medical things. *IEEE Access, 10*, 57990–58004. https://doi.org/10.1109/ACCESS.2022.3179418. [CrossRef].
22. Parra, C. M., Gupta, M., & Dennehy, D. (2022). Likelihood of questioning AI-based recommendations due to perceived racial/gender bias. *IEEE Transactions on Technology and Society, 3*, 41–45. https://doi.org/10.1109/TTS.2021.3120303. [CrossRef].

23. Camajori Tedeschini, B., Savazzi, S., Stoklasa, R., Barbieri, L., Stathopoulos, I., Nicoli, M., & Serio, L. (2022). Decentralized federated learning for healthcare networks: A case study on tumor segmentation. *IEEE Access, 10*, 8693–8708. https://doi.org/10.1109/ACCESS.2022.3141913. [CrossRef].
24. Patel, V. (2018). A framework for secure and decentralized sharing of medical imaging data via Blockchain consensus. *Health Informatics Journal.*
25. Idrees, S. M., Nowostawski, M., Jameel, R., & Mourya, A. K. (2021). Privacy-preserving infrastructure for health information systems. In *Data protection and privacy in healthcare* (pp. 109–129). CRC Press.
26. Geigel, A. (2013). Neural network Trojan. *Journal of Computer Security, 21*(2), 191–232. https://doi.org/10.3233/JCS-2012-0460
27. Zhang, J., Xue, N., & Huang, X. (2017). A secure system for pervasive social network-based healthcare. *IEEE Access, 4*, 9239–9250.
28. Magyar, G. Blockchain: Solving the privacy and research availability tradeoff for EHR data: A new disruptive technology in health data management. In *Proceedings of the 2017 IEEE 30th Neumann Colloquium (NC), Budapest, Hungary*, 24–25 November 2017; pp. 135–140. [CrossRef].
29. Weiss, M., Botha, A., Herselman, M., & Loots, G. Blockchain as an Enabler for Public MHealth Solutions in South Africa. In *Proceedings of the 2017 IST-Africa Week Conference, Windhoek, Namibia*, 31 May–2 June 2017; pp. 1–8.
30. Gordon, W. J., & Catalini, C. (2018). Blockchain Technology for Healthcare: Facilitating the transition to patient-driven interoperability. *Computational and Structural Biotechnology Journal, 16*, 224–230.
31. Ahram, T., Sargolzaei, A., Sargolzaei, S., Daniels, J., & Amaba, B. Blockchain technology innovations. In Proceedings of the 2017 IEEE Technology & Engineering Management Conference (TEMSCON), Santa Clara, CA, USA, 8–10 June 2017; pp. 137–141.
32. Reyes-Ortiz, J. A., Gonzalez-Beltran, B. A., & Gallardo-Lopez, L. Clinical decision support systems: A survey of NLP-based approaches from unstructured data. In Proceedings – International workshop on database and expert systems applications, DEXA, Feb. 2016, vol. 2016-February, pp. 163–167. https://doi.org/10.1109/DEXA.2015.47
33. Tou, H., Yao, L., Wei, Z., Zhuang, X., & Zhang, B. (2018, April). Automatic infection detection based on electronic medical records. *BMC Bioinformatics, 19*. https://doi.org/10.1186/s12859-018-2101-x
34. Kamau, G., Boore, C., Maina, E., & Njenga, S. Blockchain technology: Is this the solution to EMR interoperability and security issues in developing countries? In Proceedings of the 2018 IST-Africa week conference (IST-Africa), Gaborone, Botswana, 9–11 May 2018; pp. 1–8.
35. Bocek, T., Rodrigues, B. B., Strasser, T., & Stiller, B. Blockchains everywhere—A use-case of blockchains in the pharma supply-chain. In Proceedings of the 2017 IFIP/IEEE symposium on integrated network and service management (IM), Lisbon, Portugal, 8–12 May 2017; pp. 772–777.
36. Uddin, M. A., Stranieri, A., Gondal, I., & Balasubramanian, V. (2018). Continuous patient monitoring with a patient centric agent: A block architecture. *IEEE Access, 6*, 32700–32726.
37. Lee, S., Mohr, N. M., Nicholas Street, W., & Nadkarni, P. (2019, March 1). Machine learning in relation to emergency medicine clinical and operational scenarios: An overview. *Western Journal of Emergency Medicine, 20*(2), 219–227. https://doi.org/10.5811/westjem.2019.1.41244
38. Xu, J. et al. (2019, February 8). Translating cancer genomics into precision medicine with artificial intelligence: applications, challenges and future perspectives. *Human Genetics, 138*(2), 109–124. https://doi.org/10.1007/s00439-019-01970-5. Springer Verlag
39. Gao, J., Lanchantin, J., Soffa, M. L., & Qi, Y. (2018). Black-box generation of adversarial text sequences to evade deep learning classifiers. In *2018 IEEE Security and Privacy Workshops (SPW)* (pp. 50–56). https://doi.org/10.1109/SPW.2018.00016

40. Zhao, H., Bai, P., Peng, Y., & Xu, R. (2018). Efficient key management scheme for health blockchain. *CAAI Transactions on Intelligence Technology, 3*, 114–118.
41. Griggs, K. N., Ossipova, O., Kohlios, C. P., Baccarini, A. N., Howson, E. A., & Hayajneh, T. (2018). Healthcare blockchain system using smart contracts for secure automated remote patient monitoring. *Journal of Medical Systems, 42*, 130.
42. Patnaik, M., & Mishra, S. (2022). Indoor positioning system assisted big data analytics in smart healthcare. In *Connected e-health: Integrated IoT and cloud computing* (pp. 393–415). Springer International Publishing.
43. Esteva, A., et al. (2021, December 1). Deep learning-enabled medical computer vision. *npj Digital Medicine, 4*(1). https://doi.org/10.1038/s41746-020-00376-2
44. Khemasuwan, D., Sorensen, J. S., & Colt, H. G. (2020, September). Artificial intelligence in pulmonary medicine: Computer vision, predictive model and covid-19. *European Respiratory Review, 29*(157), 1–16. https://doi.org/10.1183/16000617.0181-2020
45. Manocha, A., & Singh, R. (2019, November). Computer vision based working environment monitoring to analyze Generalized Anxiety Disorder (GAD). *Multimedia Tools and Applications, 78*(21), 30457–30484. https://doi.org/10.1007/s11042-019-7700-7
46. Nugent, T., Upton, D., & Cimpoesu, M. (2016). Improving data transparency in clinical trials using blockchain smart contracts. *F1000Research, 5*, 2541.
47. Zhao, H., Zhang, Y., Peng, Y., & Xu, R. Lightweight backup and efficient recovery scheme for health blockchain keys. In Proceedings of the 2017 IEEE 13th international symposium on autonomous decentralized system (ISADS), Bangkok, Thailand, 22–24 March 2017; pp. 229–234.
48. Fan, K., Wang, S., Ren, Y., Li, H., & Yang, Y. (2018). MedBlock: Efficient and secure medical data sharing via blockchain. *Journal of Medical Systems, 42*, 136.
49. Choi, J., Hui, J. Z., Spain, D., Su, Y. S., Cheng, C. T., & Liao, C. H. (2021, April). Practical computer vision application to detect hip fractures on pelvic X-rays: A bi-institutional study. *Trauma Surgery and Acute Care Open, 6*(1). https://doi.org/10.1136/tsaco-2021-000705
50. Horie, Y., et al. (2019, January). Diagnostic outcomes of esophageal cancer by artificial intelligence using convolutional neural networks. *Gastrointestinal Endoscopy, 89*(1), 25–32. https://doi.org/10.1016/j.gie.2018.07.037
51. Chilamkurthy, S., et al. (2018, December). Deep learning algorithms for detection of critical findings in head CT scans: A retrospective study. *The Lancet, 392*(10162), 2388–2396. https://doi.org/10.1016/S0140-6736(18)31645-3
52. Chen, X., Yao, L., Zhou, T., Dong, J., & Zhang, Y. (2021, May). Momentum contrastive learning for few-shot COVID19 diagnosis from chest CT images. *Pattern Recognition, 113*. https://doi.org/10.1016/j.patcog.2021.107826
53. Fang, B., Mei, G., Yuan, X., Wang, L., Wang, Z., & Wang, J. (2021, May). Visual SLAM for robot navigation in a healthcare facility. *Pattern Recognition, 113*. https://doi.org/10.1016/j.patcog.2021.107822
54. Goodfellow, I. J., Shlens, J., & Szegedy, C. *Explaining and harnessing adversarial examples* [Online]. Available: https://github.com/lisa-lab/pylearn2/tree/master/pylearn2/scripts/
55. Kurakin, A., GoodfellowI., & Bengio, S. (2017, July). *Adversarial examples in the physical world* [Online]. Available: http://arxiv.org/abs/1607.02533
56. F. Tramèr et al. *Ensemble adversarial training: Attacks and defenses*.
57. Tiron, R., et al. (2020, August). Screening for obstructive sleep apnea with novel hybrid acoustic smartphone app technology. *Journal of Thoracic Disease, 12*(8), 4476–4495. https://doi.org/10.21037/jtd20-804
58. Grzywalski, T., et al. (2019). Practical implementation of artificial intelligence algorithms in pulmonary auscultation examination. *European Journal of Pediatrics*. https://doi.org/10.1007/s00431-019-03363-2
59. Ramesh, V., Vatanparvar, K., Nemati, E., Nathan, V., Rahman, M. M., & Kuang, J. (2020). CoughGAN: Generating synthetic coughs that improve respiratory disease classification. *Annual International Conference of the IEEE Engineering in Medicine & Biology Society*, 5682–5688. https://doi.org/10.1109/EMBC44109.2020.9175597

60. Sirajus, S. M., et al. (2019). Harnessing the power of deep learning methods in healthcare: Neonatal pain assessment from crying sound. In *2019 IEEE Healthcare Innovations and Point of Care Technologies, (HI-POCT)*. https://doi.org/10.1109/hi-poct45284.2019.8962827
61. Alzantot, M., Balaji, B., & Srivastava, M. *Did you hear that? Adversarial examples against automatic speech recognition.*
62. Carlini, N., & Wagner, D. (2018). Audio adversarial examples: Targeted attacks on speech-to-text. *IEEE Security and Privacy Workshops (SPW)*. https://doi.org/10.1109/SPW.2018.00009

Chapter 3
Decentralized Key Management for Digital Identity Wallets

Abylay Satybaldy, Anushka Subedi, and Sheikh Mohammad Idrees

1 Introduction

Identity management is a key component of information flow, provenance, information protection, and it acts as an enabler for many digital services. Currently, identity data and credentials are increasingly centralized by organizations and some larger corporations. Centralizing personal data in databases and data silos is problematic for various reasons, including security risks, data lock-ins, personal tracking, and targeted advertising for political influence, for instance.

Security design flaws in traditional systems pose a significant risk of data breaches. Recent examples of such breaches include incidents at Equifax [1], Cambridge Analytica [2], and First American Financial [3], where the identity information of millions was exposed. Additionally, users' identity data is fragmented across multiple providers. As a result, users must establish and manage a myriad of accounts, IDs, and passwords to interact with numerous repositories, service providers, and verifiers. The lack of sufficient privacy controls and transparency regarding how user identity data is generated, managed, and shared by third parties is another significant concern.

These challenges have spurred both industry and academia to seek innovative approaches for managing digital identity information and cryptographic secrets. Self-sovereign identity represents a novel architecture for privacy-preserving and user-centric identity management. The concept of SSI was introduced by Christopher Allen in 2016. He defined the ten principles of the SSI model [4]. SSI aims to eliminate any single point of dependency, enabling individuals to assume ownership of their digital identities. In the SSI model, there is no central authority; users maintain

A. Satybaldy · A. Subedi · S. M. Idrees (✉)
Department of Computer Science, Norwegian University of Science and Technology (NTNU), Trondheim, Norway
e-mail: sheikh.m.idrees@ntnu.no

S. M. Idrees, M. Nowostawski (eds.), *Blockchain Transformations*, Signals and Communication Technology, https://doi.org/10.1007/978-3-031-49593-9_3

their digital keys and have full control over their personal information. Typically, this information is housed in a digital identity wallet on a user's mobile device. Within the context of SSI, a digital wallet is both a software application and encrypted database that stores credentials, keys, and other essentials required for the SSI framework to function. This wallet empowers users with data sovereignty, full control, and data portability. Users can establish and engage in trusted interactions with third parties. The digital identity wallet encompasses public-private key pairs, facilitating the signing of transactions, statements, credentials, documents, or claims. Every participant in the SSI ecosystem (the issuer, holder, and verifier) requires a digital wallet to issue, maintain, and verify credentials, as depicted in Fig. 3.1.

The data stored in the wallet is safeguarded through robust encryption, a secure computing unit within the mobile device itself (i.e., Trusted Execution Environment), and user-specific knowledge (like passwords) or biometrics. While data can be delegated and managed by a third-party service, ideally, users should have the autonomy to select their service providers. Although delegation models might be more convenient, our research focuses on instances where users directly store identity data on their mobile devices.

Control over a user's digital keys, as well as other contents of the digital wallet such as credentials, is arguably the most critical element of the SSI architecture. While traditional identity management models offer key management protocols reliant on trusted third parties, in the SSI model, the responsibility of key management falls on the identity owners themselves. Given that users often lose passwords and mobile devices, placing the onus on non-technical users to safeguard credentials introduces significant risk. Therefore, for SSI to emerge as a widely adopted

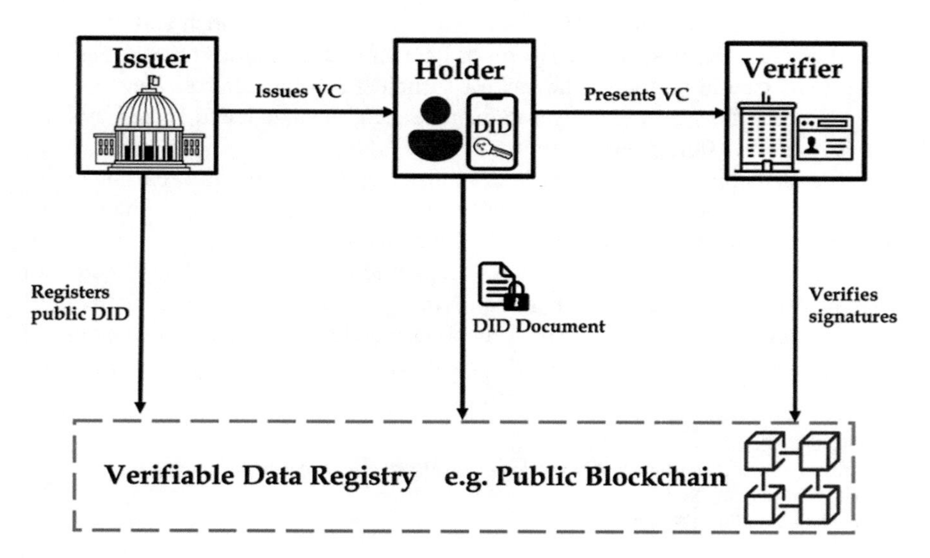

Fig. 3.1 SSI model

solution, it requires innovative decentralized key management services capable of ensuring secure and practical key distribution, verification, and recovery.

In this chapter, we present a decentralized data backup method grounded in social sharing, utilizing Shamir's Secret Sharing (SSS) algorithm [5]. As a proof of concept, we developed an application where a digital wallet's secret key is entrusted to several individuals within a user's social circle. In the event of a loss, the user can reconstruct the contents of their digital identity wallet with the assistance of these trusted individuals, or "guardians."

The remainder of this chapter is structured as follows: Sect. 2 delves into the background information on SSI, digital wallets, and Shamir's algorithm. Section 3 showcases related work in the realm of decentralized key management. In Sect. 4, we present our approach and the proposed key recovery model. Section 5 offers a comprehensive overview of our social key recovery scheme's implementation. Conclusions are drawn in Sect. 6.

2 Background

2.1 Self-Sovereign Identity

SSI is defined as a lasting identity that is owned and controlled by the individual or entity it belongs to, without dependence on any external authority and devoid of the risk of being revoked. Achieving this not only necessitates interoperability of a user's identity across various platforms (with the user's consent) but also genuine user control over that digital identity and complete autonomy. For an identity to be self-sovereign, it must be transferable and not restricted to a specific site, provider, or location. Such portability can be facilitated by an ecosystem that aids in acquiring and recording attributes, and in disseminating trust among entities using these identities.

SSI is underpinned by technologies that draw from foundational concepts in identity management, distributed computing, blockchain, and cryptography. SSI solutions should adhere to open standards and protocols such as verifiable credentials (VCs) [6] and decentralized identifiers (DIDs) [7] established by W3C. Below are brief descriptions of the architectural components:

A *digital identity wallet* offers users an interface to store, manage, and secure cryptographic keys, secrets, and other confidential data. Within the SSI framework, operating these digital wallets requires specific software termed as a *digital agent*. This agent functions as a software guardian, ensuring that only the wallet's controller (usually the identity owner) accesses the stored credentials and cryptographic keys. The agent also facilitates secure connections and credential exchanges. Furthermore, it manages third-party interactions, helping the user maintain control over their wallet.

The *decentralized identifier* (DID) represents a novel identifier generated independently of centralized authorities, and it is tailored to validate control over the DID using cryptographic validations. DIDs can verify the provenance and ownership of attested identity data by checking the proof attached to the assertion. A DID is paired with a private and public key upon creation. Every DID is linked with a DID document detailing the DID owner, which includes public keys for authentication and proof of association. This document may also list other attributes or claims about the owner and specify which entities can modify the DID document. Typically represented in JSON-LD [8], these documents can also be conveyed using other data formats.

A *verifiable credential* (VC) can embody the same data that physical credentials do. Being tamper-evident and having a cryptographically verifiable authorship, VCs are often deemed more reliable than their tangible counterparts. Generally, every VC contains claims about its subject. In the SSI model, we have three major actors: issuer, holder, and verifier.

Issuer formulates a claim about a specific subject, connects it with the subject's DID and DID document, and converts this claim into a verifiable credential for transmission to a holder. Issuers can range from corporations and governments to individuals.

Verifier accepts claims in the form of VCs from other entities to grant them access to protected resources. Verifiers can cryptographically ascertain the legitimacy of the shared information without liaising directly with the initial issuer. The digital signature of the issuer is authenticated using a DID coupled with a distributed ledger or another decentralized network.

Holder (a person, organization, or object) receives the credential and stores it in their digital wallet. Holders can present claim proofs from one or more credentials upon request from verifiers, as illustrated in Fig. 3.1. Typically, the holder is the subject of the credentials they possess, though this is not always the case.

In the SSI paradigm, *verifiable data registry* (VDR) oversees the creation and verification of identifiers, keys, and other pertinent data such as verifiable credentials, revocation lists, and issuer public keys. Examples of such registries include blockchains and decentralized databases. When a DID and its initial public key are "recorded" in a blockchain via a digitally signed transaction, the blockchain emerges as the trust anchor for the DID. This mandates that verifiers consult the blockchain to confirm the current public key and any associated DID document content. In essence, trust in the consensus algorithm and the specific ledger's operation is imperative. Given the reliability of established public blockchains like Bitcoin and Ethereum, and the robust methods to authenticate lookups from these ledgers, they are broadly perceived as robust trust anchors.

2.2 Shamir's Secret Sharing

Shamir's Secret Sharing (SSS) scheme is an algorithm first introduced in 1979 by the esteemed Israeli cryptographer, Adi Shamir. The SSS is recognized as information-theoretically secure and has been employed in cryptographic contexts and other high-security applications [9].

Shamir's scheme, based on standard Lagrange polynomial interpolation, permits a secret to be divided into multiple shares. Intriguingly, reconstructing the original secret requires only a subset of these shares. In a variety of sectors, both researchers and businesses are harnessing this algorithm to incorporate decentralized sharing into their respective solutions, services, and products. The SSS scheme has been employed to encrypt data at the field level, enabling storage in cloud databases without undermining data privacy [10].

In [11], researchers proposed a reversible data hiding scheme anchored in Shamir's Secret Sharing and designed to verify rightful ownership in encrypted databases. This method conceals secret information by dispersing it into numerous random-looking shares. Furthermore, SSS has been utilized for cryptographic key sharding and recovery. For instance, it is used to divide PGP Desktop keys into multiple segments, which can later be merged when necessary [12].

Hashicorp's Vault [13] integrates SSS into its unsealing process, which retrieves and decrypts secrets from cloud storage. Instead of conferring the unseal key as a singular entity to an operator, Vault employs Shamir's algorithm to fragment the key, thereby eliminating any single point of vulnerability.

3 Related Work

Any form of cryptographic key management is challenging for humans because digital keys are essentially strings of bits that must be meticulously safeguarded. If they are lost, stolen, or corrupted, they can be virtually irreplaceable. Standards and protocols for conventional key management are well established, with notable publications from NIST and the Key Management Interoperability Protocol (KMIP) from OASIS [14].

The transition to decentralized key management signifies a move from centralized roots of trust to algorithmic or self-certifying roots of trust. Both of these approaches eliminate the need to place trust in humans or organizational assertions of new or rotated keys. However, this shift introduces new key management responsibilities that now rest squarely on the shoulders of self-sovereign individuals. With SSI, there is no overarching authority to consult. This necessitates new open standards and protocols that allow digital wallets to be user-friendly and compatible across vendors, devices, systems, and networks.

Several organizations and companies are already addressing the challenges of decentralized key management. The increasing interest in decentralized identity prompted the US Department of Homeland Security (DHS) to award a research contract on decentralized key management to SSI vendor Evernym in 2017 [15]. Evernym convened a team of cryptographic engineers and key management specialists to draft a document titled "Decentralized Key Management System (DKMS): Design and Architecture." This was subsequently published as part of the Hyperledger Indy project at the Linux Foundation [16]. DKMS outlines the design prerequisites and offers guidelines for creating decentralized key management solutions. The Decentralized Identity Foundation (DIF) also has a Wallet Security working group dedicated to formulating guidelines and setting security standards relevant to identity wallet architectures, such as key management, credential storage, credential exchange, backup, recovery, and wallet portability [17].

Below is an overview of the current key recovery solutions being used, implemented, or considered:

Offline recovery. A mnemonic (seed phrase) is a human-readable sequence of 12 to 24 words that encodes the wallet's root private key. Users are responsible for backing up their recovery mnemonic and ensuring its safety so they can regenerate their private key if necessary. This recovery method is standard for cryptocurrency wallets and some identity wallets at present. While mnemonics are a familiar mechanism among decentralized application users, their security is only as robust as where they are stored. For instance, when noted on paper, they are vulnerable to fire, floods, theft, and deterioration.

Biometrics. Key management based on biometric traits has garnered significant attention in the research community [18]. Biometric traits, being portable and largely unique, can serve as seed values for generating cryptographic keys. The drawback of this method is that once a person's biometric data becomes public, it can no longer secure an account since one cannot alter a fingerprint as easily as changing a password or switching an account. Another limitation of biometrics is the potential inconsistency across different fingerprint sensors – for example, a minor injury might cause recognition issues [19].

Social recovery. This recovery method involves trusted entities, known as "trustees," who store recovery data on behalf of the identity owner, typically in the trustee's own wallet. While no current digital identity wallets support social key recovery, some cryptocurrency wallets, including Argent and Loopring, have implemented this feature [20, 21].

4 Methodology

The main objective of our proposed solution is to make a private key recoverable by splitting it into chunks, distributing them to a set of trustees, and, when the original is lost, recovering the full key from the full set or a subset of the chunks. Initially, the user of the digital identity wallet generates their private key, which allows them to have full control over their personal data and identity information. However, the original private key should be stored securely so that the user can recover it later in case of loss or theft. The core of the scheme is based on Shamir's Secret Sharing algorithm, which divides the private key into parts called shares. These shares are then distributed to a group of people. For recovery, these parts are combined to reconstruct the original secret key. In our case, recovering the master key allows for the restoration of the wallet. An essential feature of the scheme is that the reconstruction does not require all the participants, but a threshold number of participants decided upon during the division of the original key. Any combination of shares less than the necessary threshold conveys absolutely no information about the secret. Additionally, to protect against possible collusion among trustees, the private key can be password-encrypted so that only the original owner (who knows the passphrase) can access the actual reconstructed key.

In the following paragraph, we describe the creation of the secure private key and the recovery of the original private key, as illustrated in Fig. 3.2.

The digital wallet initially generates a cryptographic private key K_{prv}. Given a set of T trustees $t_1...t_n$, Algorithm 1 divides the key into a set of shares $S[i]$ for n trustees using the SSS split function. The user can set the threshold value r, which is the minimum number of participants required to reconstruct the secret. In the subsequent step, the wallet creates a data package to be sent to each participant. For every trustee i, the package consists of the key share $S[i]$, the total number of trustees n, the threshold value r, and the digital signature generated based on the aforementioned items. The wallet then securely shares each encrypted package $Enc(P[i], K_{pub}[i])$ with trustees, where $K_{pub}[i]$ is the public key of trustee i.

Algorithm 2 outlines the recovery process of the original private key. The algorithm first checks if the threshold value is satisfied. If there are enough shares ($|T| \geqq r$), it iterates through those tuples, decrypts the secure key ($K_{sec}[i]$), and produces the package containing the key share. Finally, it passes the decrypted chunks through the SSS combine function, which yields the original private key K_{prv}.

Algorithm 1: Create
Require: a set of **T** trustees **$t_1...t_n$**
1: **Generate** private key **K_{prv}**
2: **Define** threshold value **r**
3: **for** i=0 **to** n **do**
4: S[i] = **SSS_{fn}**(K_{prv})
5: P[i] = (S[i], n, r)
6: K_{sec}[i] = **Enc**(P[i], K_{pub}[i])
7: **end for**
8: **return** K_{sec}

Algorithm 2: Recover
Require: number of trustees |T| ≥ r
2: **if** |T| ≥ r
3: **for** i=0 **to** |T| **do**
4: P[i] = **Dec**(K_{sec}[i])
5: S[i] <- P[i]
4: **end for**
5: K_{prv} = **SSS_{Fn}**(S)
6: **end if**
7: **return** K_{prv}

Fig. 3.2 Key generation and recovery algorithm

5 Implementation

5.1 Key Generation

In order to design a working proof of concept, it was crucial for the key generation component to be executed by the digital identity wallet integrated into our proposed solution. This integration would demonstrate that the proposed scheme is compatible with other existing digital identity wallets. For this purpose, we analyzed several identity wallet frameworks provided by Trinsic [22], Veramo [23], and SpruceID [24]. All these frameworks are open source and do not have a key recovery feature, as shown in Table 3.1. Due to its flexible plugin system, compatibility with W3C and DIF standards, and JavaScript support, we chose the Veramo framework for our use case.

First, we bootstrapped our agent using Veramo APIs. This was achieved by initializing the npm package "@veramo/core." The Veramo agent offers a common interface for plugins to extend their functionality, as depicted in Fig. 3.3. Our key recovery feature can be integrated as a custom plugin into the Veramo framework. The agent is also the primary class and, when instantiated, organizes both core and custom plugins and manages the core event system. After setting up the agent, the *did-manager* and *key-manager* core plugins were employed to create a DID and store the private key, respectively. SQLite was the chosen database for storing the key, and Node was utilized as the development framework.

Table 3.1 Comparative analysis of SSI wallet frameworks

Features	Trinsic	Veramo	Spruce
Open source	Yes	Yes	Yes
W3C DIDs and VCs	Yes	Yes	Yes
Key recovery support	No	No	No
Supported DID methods	Did:Sov, did:Peer, did:Key, did:Web	Did:Ethr, did:Web, did:Key	Did:Web, did:Key, did:Tezos, did-ethr, did:Sol
Blockchains	Hyperledger Indy, Sovrin	Ethereum	Tezos, Solana, Ethereum
Proof data formats	JSON-LD	JSON-LD	JSON-LD
Proof types (signatures)	Ed25519, CL signatures	Secp256k1, Ed25519, RSASignature2018	Ed25519, RSASignature2018, Eip712Signature2021, JsonWebSignature2020
Zero-knowledge proofs	Yes	No	No
Framework language	.Net	JavaScript	Rust

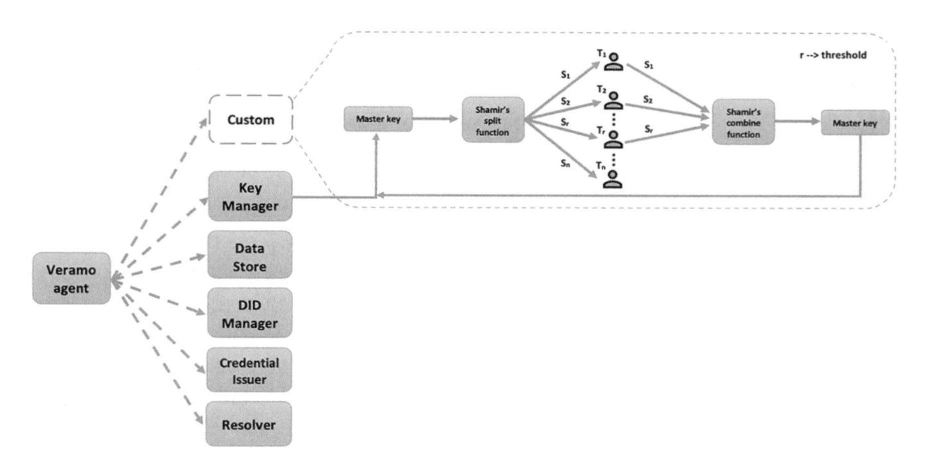

Fig. 3.3 Key recovery protocol integrated to Veramo key management plugin

5.2 *Key Sharing*

In our implementation, we utilized Shamir's Secret Sharing scheme to divide a private key, generated by the identity wallet, into segments referred to as shares. The number of shares is flexible; that is, a user can split the key into two or more shares. These shares are distributed to a group of individuals designated as trustees by the user. These segments of the secret can be combined to reconstruct the original secret. Notably, one does not need all the shares to reconstruct the key. Furthermore, a single share reveals no information about the secret. A specific number of shares, less than the total and referred to as the threshold number, is necessary for reconstruction. This threshold is determined when creating the shares. The concept of consolidating the threshold number of shares to recover the key mitigates decryption failures when one or several parties are unavailable.

As an initial proof of concept, we developed a web application in *ReactJS* where users could add friends and family members. Basic validations, such as preventing the same individual from being added as both a family member and a friend or adding a user multiple times, were implemented. Once these fields were filled, the private key generated from the DID was divided into shares using the *shamirs-secret-sharing* library.[1]

[1] SSS library: https://www.npmjs.com/package/shamirs-secret-sharing

5.3 *Key Recovery*

According to the SSS, all secrets carry equal weight, and once a sufficient threshold is reached, the secret can be reconstructed. However, in real-world scenarios, trust is not uniformly distributed across different social circles, such as friends, business acquaintances, and family. For instance, a family member and a business partner might not hold the same level of trust for some individuals. Consequently, we introduced a social key recovery scheme with a priority setting system. As a proof of concept, we implemented two categories with fixed priorities: family and friends. Family members are allocated two shares of the key, while friends receive one share each. Therefore, if three shares are needed for wallet recovery, just one family member and one friend would suffice to reconstruct the secret. The git repository for this implementation is available on GitHub.[2] Instructions in the README.md file within the repository guide users on how to run the system locally. It successfully generates shares for distribution among the designated trustees and recovers the secret when sufficient shares are provided, without any failures.

6 Discussion

Our proposed model outlines an architecture in which a digital identity wallet user can recover their private key without requiring a physical backup. These methods are complementary and should be combined for optimal security. Recovering the private key enables the regeneration of all subkeys and the restoration of the wallet. There are established techniques, such as secure hierarchical deterministic (HD) key generation, which can use a single master seed to generate numerous key pairs [25]. However, if you lose your mobile device, you might still forfeit all your previous credentials and connections stored in the mobile wallet, even after recovering the private key. Therefore, users are encouraged to back up encrypted wallet data on cloud-based services.

The private key can be distributed among trusted individuals within a user's social circle. Collaboration among fewer than the threshold number of trustees results in unsuccessful reconstruction of the secret keys. As the number of trustees and the threshold increase, the system's security is enhanced, but its efficiency and usability might diminish. For a user's digital wallet to encrypt each package with a key share, the wallet should possess access to the receiving party's public key. Our model presumes a decentralized trust model based on distributed ledger technology, and DIDs facilitate the ability to look up public DIDs and establish connections. In the proposed protocol, a secure and encrypted communication channel among the digital wallets is essential. This can be established using the DIDComm protocol [26].

[2] Social key recovery scheme: https://github.com/abylays/Key-recovery-scheme

A challenge with social key recovery is the potential for the group of trustees to collude and gain unauthorized access to the account without the owner's consent. Therefore, in our model, group members are unaware of the other participants. A key recovery mechanism must be implemented with caution. If not deployed correctly, it could compromise security. Our scheme strikes a balance between security – having an encrypted version of the original private key safe for storage in less secure environments – and usability – being able to retrieve the original private key after extended periods without relying on easily forgotten passphrases or hardware security modules.

The use case implementation showcases how our approach can be seamlessly integrated into a real-world identity wallet/agent and how the registration and recovery processes are executed. Additionally, we introduced a social key recovery scheme with a priority setting system. As a result, our system provides users with a more intuitive and user-friendly process.

7 Conclusion

In decentralized identity management, recovery is crucial since identity owners lack a higher authority to which they can turn for assistance. In this chapter, we introduced a decentralized key backup and recovery model for SSI wallets based on Shamir's Secret Sharing algorithm. As a proof of concept, we developed an application where the secret key from a digital wallet is distributed to several individuals from the user's social circle. This ensures that if the user loses access, they can reconstruct the contents of their digital identity wallet with the assistance of their guardians.

For future work, we plan to fully develop our digital wallet software, incorporating a default key recovery feature. We aim to conduct comprehensive security analyses and carry out usability tests with actual users. Additionally, we are interested in exploring innovative methods that could enhance our secret sharing algorithm and bolster the security of our model.

References

1. Berghel, H. (2017). Equifax and the latest round of identity theft roulette. *Computer, 50*(12), 72–76.
2. Isaak, J., & Hanna, M. J. (2018). User data privacy: Facebook, Cambridge analytica, and privacy protection. *Computer, 51*(8), 56–59.
3. Dellinger, A. (2022). *Understanding the first American financial data leak: How did it happen and what does it mean?* Available at https://bit.ly/3CfTvlC. Accessed 11 June 2023
4. Allen, C. (2016). *The path to self-sovereign identity*. Available at http://www.lifewithalacrity.com/2016/04/the-path-to-self-sovereign-identity.html. Accessed 22 June 2023.
5. Shamir, A. (1979). How to share a secret. *Communications of the ACM, 22*(11), 612–613.

6. W3C Credentials Community Group. (2022). *Verifiable credentials data model v1.1*. Available at https://www.w3.org/TR/vc-data-model/. Accessed 22 Aug 2023
7. Reed, D., Sporny, M., Longley, D., Allen, C., Grant, R., Sabadello, M., & Holt, J. (2022). *Decentralized identifiers (DIDs) v1. 0. W3C recommendation*.
8. Kellogg, G., Champin, P. A., & Longley, D. (2019). *JSON-LD 1.1–A JSON-based serialization for linked data (W3C working draft)* (Doctoral dissertation, W3C).
9. Dawson, E., & Donovan, D. (1994). The breadth of Shamir's secret-sharing scheme. *Computers & Security, 13*(1), 69–78.
10. Tawakol, A. M. (2016). *Using Shamir's secret sharing scheme and symmetric key encryption to achieve data privacy in databases*.
11. Singh, P., & Raman, B. (2018). Reversible data hiding based on Shamir's secret sharing for color images over cloud. *Information Sciences, 422*, 77–97.
12. Broadcom. *How to split and rejoin PGP desktop 8.x keys*. Available at https://knowledge.broadcom.com/external/article/180108/how-to-split-and-rejoin-pgp-desktop-8x-k.html. Accessed 30 July 2023
13. HashiCorp. *Vault*. Available at https://www.vaultproject.io/docs/concepts/seal. Accessed 30 July 2023
14. OASIS. *Key management interoperability protocol (KMIP)*. Available at https://www.oasis-open.org/committees/kmip. Accessed 30 Aug 2023
15. DHS. *S&T awards 749k to Evernym for decentralized key management research and development*. Available at www.dhs.gov/science-and-technology/news/2017/07/20/news-release-dhs-st-awards-749k-evernym-decentralized-key. Accessed 13 Aug 2023
16. Hyperledger Foundation. *DKMS (decentralized key management system) design and architecture v4*. Available at https://github.com/hyperledger/aries-rfcs/blob/main/concepts/0051-dkms/dkms-v4.md. Accessed 20 June 2023.
17. Decentralized Identity Foundation (DIF). (2022). *Wallet Security Group*. Available at https://identity.foundation/working-groups/wallet-security.html, Accessed 27 May 2023.
18. Chen, B., & Chandran, V. (2007, December). Biometric based cryptographic key generation from faces. In *9th biennial conference of the Australian pattern recognition society on digital image computing techniques and applications (DICTA 2007)* (pp. 394–401). IEEE.
19. Bhatega, A., & Sharma, K. (2014, December). Secure cancelable fingerprint key generation. In *2014 6th IEEE Power India International Conference (PIICON)* (pp. 1–4). IEEE.
20. Loopring Protocol. (2023). *Loopring smart wallet with social recovery*. Available at https://medium.loopring.io/?gi=248a43681b24. Accessed 22 Aug 2023
21. Argent Labs. (2022). *How to recover my wallet with guardians*. Available at https://support.argent.xyz/hc/en-us/articles/360022631412-About-wallet-recovery, Accessed 7 June 2023
22. Trinsic. A full-stack SSI platform. Available at https://trinsic.id/. Accessed 23 Aug 2023.
23. Veramo. Available at https://veramo.io/. Accessed 13 Aug 2023.
24. SpruceID. *Your keys, your data*. Available at www.spruceid.com/. Accessed 18 Aug 2023
25. Allen, C., & Appelcline, S. *Hierarchical deterministic keys: Bip32 and beyond*. Available at https://github.com/WebOfTrustInfo/rwot1-sf/blob/master/topics-and-advance-readings/hierarchical-deterministic-keys–bip32-and-beyond.md. Accessed 13 Aug 2021
26. Curren, S., & Looker, T. & Terbu, O. *DIDComm messaging v2.x editor's draft*. Available at https://identity.foundation/didcomm-messaging/spec/. Accessed 23 June 2023.

Chapter 4
Towards Blockchain Driven Solution for Remote Healthcare Service: An Analytical Study

Siddhant Prateek Mahanayak, Barat Nikhita, and Sushruta Mishra

1 Introduction

Blockchain is an innovative decentralized method of keeping records that allows for secure, open, and unchangeable recording of transactions [1]. It was initially developed as the underlying technology behind Bitcoin, a cryptocurrency, in 2008 but has since evolved to encompass many applications beyond just cryptocurrencies [2]. At its essence, Blockchain is a distributed database where multiple computers, called nodes, share a copy of the database. A consensus mechanism is used to add new data to the database to ensure the network's integrity and security. Once data is added to the database, it becomes immutable and tamper-proof, and changes to the database cannot be made without the agreement of the entire network. By improving the security, privacy, and interoperability of medical data and enabling safe data exchange and effective data management across many healthcare stakeholders and providers, the field of telemedicine could undergo a significant revolution thanks to the transformative capabilities of blockchain technology.

A key feature of blockchain technology is decentralization, where several nodes participate in the network, and each one has the same degree of power to administer the ledger, validate transactions, and make decisions [3, 4]. This distributed nature of Blockchain offers several advantages, such as enhanced transparency and immutability. Since each node possesses a replica of the ledger, any modifications require the consensus of most other nodes in the network. This feature makes sure that the data on blockchain are almost immune to corruption [5], thus making it a highly secure and reliable technology (Fig. 4.1).

S. P. Mahanayak · B. Nikhita · S. Mishra (✉)
Kalinga Institute of Industrial Technology, Deemed to be University,
Bhubaneswar, Odisha, India
e-mail: sushruta.mishrafcs@kiit.ac.in

S. M. Idrees, M. Nowostawski (eds.), *Blockchain Transformations*, Signals and Communication Technology, https://doi.org/10.1007/978-3-031-49593-9_4

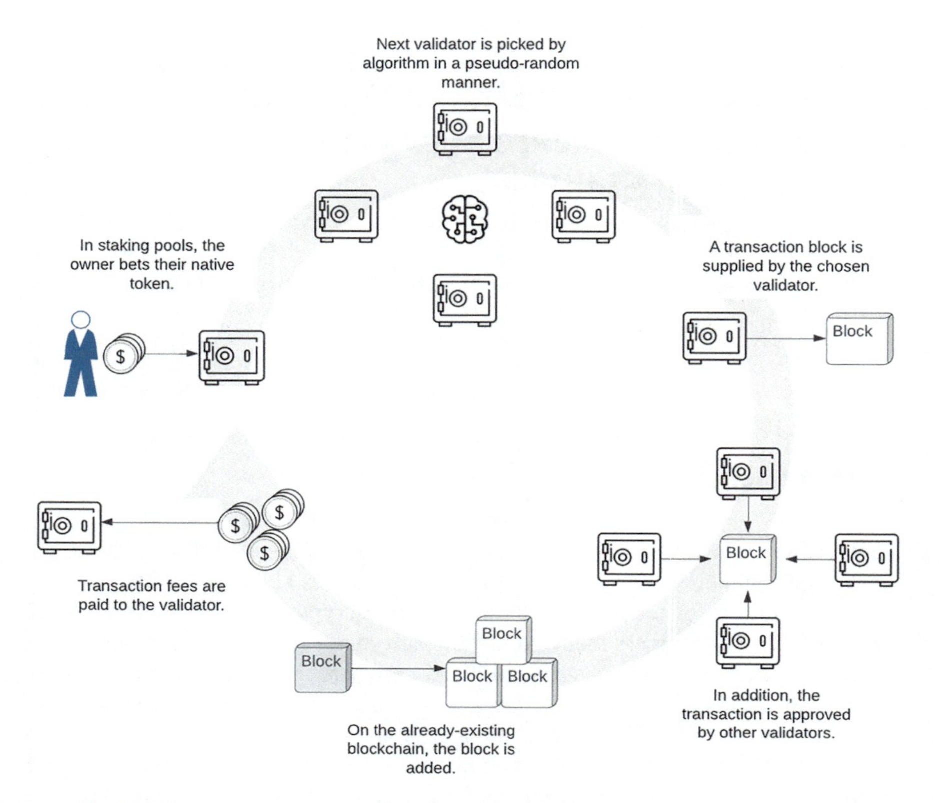

Fig. 4.1 Proof of Stake

In blockchain technology, smart contracts are a specific form of contract that function within a blockchain network and are coded to encompass the conditions of an agreement between a purchaser and vendor [6, 7]. These contracts are self-executing and comprise lines of code that can automatically trigger specific actions based on pre-set rules and conditions. This means that a smart contract can facilitate the exchange of goods or services between parties without the need for intermediaries or manual intervention. For example, an intelligent agreement can be coded to automatically transfer funds to the seller as soon as the buyer confirms the receipt of the product or service. This simplifies and automates the entire process of the transaction. It is a crucial concept in blockchain technology, which involves transforming a piece of data, such as a block or transaction, into a fixed-length alphanumeric string known as a hash [8] (Fig. 4.2).

This process is irreversible, and each original data produces a unique hash. Hashing algorithms are designed to be unidirectional, making it almost impossible to reconstruct the original data from the hash. This characteristic makes hashing a valuable tool for creating a secure and unalterable record of transactions on a blockchain [8]. Decentralized blockchain networks lack a central authority for

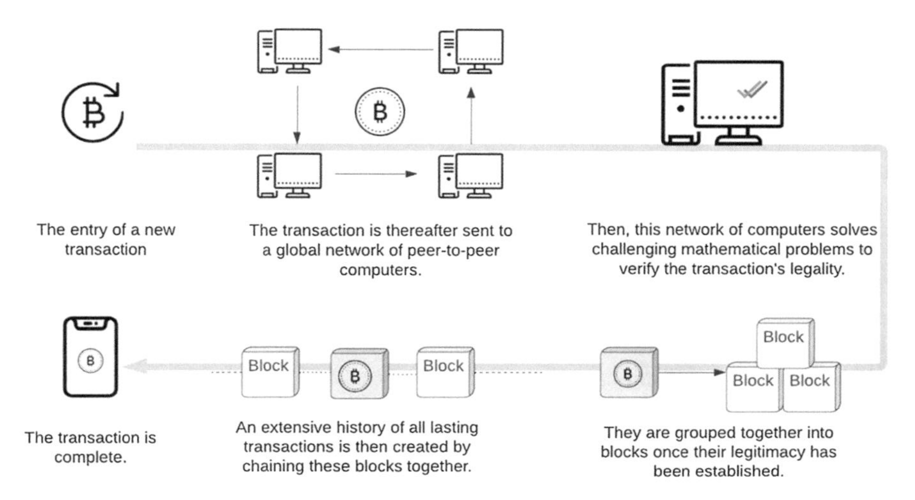

Fig. 4.2 Transaction processing

transaction verification, leading to the adoption of consensus algorithms to guarantee uniformity of the ledger across all network nodes. The blockchain community has devised multiple consensus algorithms for this purpose [9–11], including:

- Proof of Work (PoW): This mechanism involves a competition among nodes in a network to solve intricate mathematical challenges to verify transactions and generate fresh blocks [12, 13]. The initial node to solve the problem is granted with new digital currency units as an incentive. PoW is extensively employed in well-known cryptocurrencies such as Bitcoin [14, 15].
- Proof of Stake (PoS): In Fig. 4.1, this algorithm allows nodes to verify transactions and generate new blocks based on their cryptocurrency holdings within the network. The likelihood of a node being chosen to generate the next block is directly related to the quantity of cryptocurrency it owns [16, 17]. This process, known as Proof of Stake (PoS), is employed by several popular cryptocurrencies such as Ethereum and Cardano [18].
- Delegated Proof of Stake (DPoS): DPoS is a type of consensus mechanism that is utilized by specific blockchain networks to establish agreement among the various nodes within the network [19, 20]. In a Delegated Proof of Stake (DPoS) system, token holders can vote for delegates who are responsible for verifying transactions and creating fresh blocks within the Blockchain. This process is proportional to the number of tokens each voter holds, with the top candidates selected as delegates [21].
- Byzantine Fault Tolerance (BFT): BFT (Byzantine et al.) protocols have been created to guarantee the resilience of distributed systems by utilizing various methods, including redundancy, replication, and consensus algorithms [22, 23]. These methods are employed to ensure the proper operation of the system, even in cases where specific nodes fail or engage in malicious activities. In a BFT

(Byzantine et al.) system, identical data copies are stored on various nodes, and these duplicates must agree on the exact value before they can be considered valid. This agreement process guarantees that the accurate value can be determined, even if some nodes fail or supply incorrect data, thus safeguarding the system's integrity and precision [23].

2 Emergence of Remote Healthcare

Remote healthcare, which is also called telehealth or telemedicine, is the delivery of medical services to patients using digital communication technologies like video conferencing, eliminating the requirement for face-to-face appointments [24]. This method allows healthcare professionals to diagnose, treat, and monitor patients from a distance, providing them more flexibility, convenience, and easy access to medical services without leaving their homes. Consequently, the utilization of remote healthcare has seen a rise in recent times, particularly during the COVID-19 outbreak [25], due to its ability to reduce the risk of infection transmission and potentially lower healthcare costs. Furthermore, telemedicine can enhance the availability of healthcare services, particularly for individuals living in rural or isolated regions, all the while enhancing the effectiveness of healthcare delivery. Remote healthcare has several advantages, including:

- Convenience: Remote healthcare offers the convenience of receiving medical care from one's own home, eliminating the need for patients to visit healthcare facilities, which can be especially beneficial for those living in remote or rural areas or individuals who have difficulty with mobility [26]. This mode of healthcare delivery saves patients time and effort while reducing the stress associated with travelling to medical appointments, ultimately improving patient experience and satisfaction.
- Cost-effective: Remote healthcare has the potential to be a more cost-effective alternative to traditional healthcare services, as it eliminates the expenses associated with patient travel to medical facilities and can potentially reduce hospitalization rates [27]. This approach enables healthcare providers to save on overhead costs, including staffing, equipment, and facility maintenance, while offering patients affordable healthcare options. As a result, remote healthcare can be a financially viable solution for patients, especially those who face financial constraints or are uninsured.
- Improved access: Remote healthcare can enhance access to healthcare services, particularly for patients who face barriers to receiving care, such as individuals residing in underserved areas or with limited mobility. This mode of healthcare delivery eliminates geographical and transportation-related obstacles, allowing patients to connect with healthcare providers from virtually anywhere. Furthermore, remote healthcare can address the shortage of healthcare professionals in certain areas, providing patients access to specialized medical services

and reducing wait times. By improving access to healthcare, remote healthcare can contribute to better health outcomes and overall patient well-being.

- Reduced risk of infection: Remote healthcare can lower the risk of infectious disease transmission, as patients can receive medical care without being exposed to other patients who may be carrying contagious illnesses. This mode of healthcare delivery minimizes the need for patients to physically visit healthcare facilities physically, thereby reducing the chances of infection spreading through contact with potentially infected individuals. Additionally, remote healthcare can be especially valuable during disease outbreaks, as it enables healthcare providers to continue offering medical services while minimizing the risk of disease transmission, safeguarding the health of both patients and healthcare workers.
- Increased efficiency: Remote healthcare can improve healthcare efficiency by enabling providers to see a higher volume of patients in a shorter period while reducing appointment wait times. This approach allows doctors to offer medical services remotely without being restricted by the physical constraints of traditional healthcare settings. By reducing the need for patients to travel to healthcare facilities, remote healthcare can also minimize appointment delays caused by traffic, parking, or other scheduling conflicts. Furthermore, remote healthcare can facilitate faster communication and information exchange between healthcare providers, improving coordination of care and ultimately leading to better patient outcomes. Overall, remote healthcare can streamline healthcare delivery and improve healthcare system efficiency.

In general, telemedicine can enhance the availability of healthcare services, lower expenses, and enhance effectiveness. As technological advancements continue, telemedicine will probably play a progressively significant role within the healthcare system.

3 Blockchain in Remote Healthcare

Blockchain technology has arisen as an encouraging advancement with the capacity to revolutionize the healthcare sector, specifically in remote healthcare. Blockchain technology can improve the effectiveness, safety, and openness of remote healthcare services while strengthening patient confidentiality and safeguarding data. By leveraging the decentralized and immutable nature of Blockchain, healthcare providers can ensure the secure and accurate sharing of patient data between different stakeholders while minimizing the risk of data breaches and unauthorized access [28]. Furthermore, blockchain technology can enhance the tracking and management of medical records, facilitate remote identity verification and authentication, and streamline payment and reimbursement processes, all of which contribute to a more efficient and patient-centred healthcare system. Overall, blockchain technology represents a promising avenue for advancing remote healthcare delivery and improving healthcare outcomes.

One significant benefit of incorporating blockchain technology in the healthcare sector is its ability to create a reliable and unchangeable account of transactions while ensuring security [29, 30]. This attribute can be precious in remote healthcare for upholding patient privacy and data security. By leveraging blockchain technology, healthcare providers can establish a tamper-proof repository to store patient medical records and personal information, which can only be accessed by authorized stakeholders. This measure can act as a protective measure to prevent unauthorized entry into private patient information and reduce the possibility of data breaches [31]. The utilization of blockchain technology in remote healthcare can enable transparent and verifiable patient data access, empowering patients with increased authority over their personal health information. By guaranteeing the security and confidentiality of patient data, blockchain technology can boost patients' trust and confidence in remote healthcare services, thus encouraging the broader acceptance and implementation of this method of delivering healthcare [32].

Beyond enhancing data security, blockchain technology can contribute to optimizing remote healthcare services by implementing smart contracts. By leveraging smart contracts, healthcare providers can automate various administrative tasks and reduce the need for intermediaries in healthcare transactions. To reduce the stress on healthcare providers and to increase the speed and accuracy of payment processing, smart contracts may be designed, for instance, to check insurance coverage and execute payments automatically. This may increase operational effectiveness and reduce costs, allowing healthcare practitioners to concentrate on providing patients with high-quality treatment [7, 33]. Furthermore, smart contracts can improve the clarity and trackability of healthcare transactions. This empowers patients with better insight into the payment and reimbursement procedures. Incorporating blockchain technology into remote healthcare can transform how healthcare transactions are carried out, facilitating more effective and patient-focused care [34].

Incorporating blockchain technology into remote healthcare can also yield a potential advantage by enabling greater transparency and accountability [35, 36]. Healthcare providers can create an unchangeable and open record of all transactions carried out on the network by incorporating blockchain technology [37]. This record can be leveraged for auditing and compliance purposes, facilitating the verification of regulatory compliance and ethical standards adherence by healthcare providers. Additionally, blockchain technology can establish a clear and transparent log of all patient interactions with healthcare providers, enabling patients to access a complete and unalterable history of their healthcare interactions [38]. By increasing transparency and accountability, blockchain technology can enhance patient trust and satisfaction with remote healthcare services, fostering increased adoption and utilization. Blockchain technology can provide significant advantages for remote healthcare services, elevating healthcare transactions' efficiency, security, and accountability. Moreover, blockchain technology can also enhance the accuracy and comprehensiveness of patient medical records in remote healthcare [39]. With the use of a decentralized ledger, patient medical records can be securely stored and made easily accessible to authorized healthcare providers, regardless of their physical location. This can lessen the possibility of mistakes and omissions and guarantee that patients receive the best possible treatment. Blockchain technology can play a crucial role in improving healthcare by allowing healthcare providers to access patient medical records smoothly and securely. This ensures that healthcare professionals possess comprehensive knowledge about a patient's medical background, leading to more knowledgeable and precise diagnoses and treatment strategies. Therefore, integrating blockchain technology into remote healthcare can significantly enhance the quality and safety of patient care.

Integrating blockchain technology in remote healthcare services holds immense potential for revolutionary changes in the industry. By strengthening data security, simplifying administrative procedures, promoting transparency and accountability, and enhancing the precision and comprehensiveness of patient medical records, blockchain technology can have a crucial impact on ensuring patients receive excellent care, regardless of location. As remote healthcare services gain popularity and become more widespread, utilizing blockchain technology is expected to become even more critical for maximizing the efficiency of healthcare delivery. Therefore, it is clear that the incorporation of blockchain technology in remote healthcare has the potential to revolutionize the healthcare industry and bring advantages to both patients and healthcare providers.

- The objective is to suggest a practical and secure method of storing and sharing confidential medical imaging data in remote healthcare using blockchain technology. This proposed system aims to increase the data's security and integrity while improving its scalability, efficiency, and user experience. Despite the challenges of implementing this system, such as regulatory and legal barriers and technical expertise requirements, it is a promising approach that could greatly benefit remote healthcare data management. This system could be realized and widely adopted through collaboration between healthcare providers, technology companies, and regulatory bodies, transforming how medical imaging data is managed in remote healthcare.
- The objective is to showcase the practicality of utilizing blockchain technology implemented with Hyperledger for storing and distributing confidential medical imaging data while also assessing the system's effectiveness and safety. This study investigates the system's performance and security to ensure its reliability and practicality for remote healthcare data management.
- This research aims to address the concerns related to data security, privacy, and interoperability in remote healthcare. The proposed solution involves utilizing blockchain technology. The study will investigate how Blockchain can effectively deal with these issues and enhance medical data management in remote healthcare environments. Furthermore, it will evaluate the practicality and efficacy of implementing blockchain technology in remote healthcare and explore strategies to overcome any obstacles to its adoption.

The primary goals of this research involve creating a thorough structure for conducting additional investigations into the application of blockchain technology in remote healthcare. This structure intends to recognize the possible advantages and difficulties associated with implementing blockchain technology in healthcare, offer recommendations for overcoming technical and regulatory obstacles, and propose future research avenues. Additionally, this study seeks to promote further exploration and advancement in blockchain technology and facilitate the integration of blockchain solutions within the healthcare industry.

4 Related Works

To safeguard patient privacy, preserving the privacy of delicate and individual information found within medical images is of utmost importance. Healthcare professionals are mandated to adhere to rigorous guidelines and regulations, such as HIPAA [40], to protect the confidentiality and privacy of medical data. Furthermore, it is necessary to establish access restrictions to guarantee that individuals with proper authorization can access and distribute medical images.

Researchers like Chhikara et al. [41] observed that current digital healthcare systems often have problems securely and efficiently sharing data, which can negatively impact patient outcomes and increase costs. To solve this issue, they proposed a mechanism that utilizes blockchain technology to create a decentralized and secure network for authorized healthcare providers to share data. They also explained the proposed mechanism's structure and workflow in detail, including implementing smart contracts to automate access control and data sharing. Hathaliya et al. [42] highlighted the primary difficulties in monitoring patients remotely, including privacy, data security, and interoperability concerns. To overcome these challenges, the authors propose an architecture that utilizes blockchain technology to manage patient data using smart contracts. This approach is designed to provide a secure and confidential way of collecting and storing patient data while ensuring its integrity. Neyire Deniz Sarier [43] introduced a new method for securely and privately using biometric data for authentication in healthcare systems. The proposed approach utilizes blockchain technology and smart contracts to store encrypted biometric data off-chain and manage the authentication process. Simulations were conducted to test the system's performance, and the results indicate that the proposed approach can handle a high volume of authentication requests without causing significant delays. Overall, the system presents a promising solution to the security and privacy concerns associated with biometric authentication in healthcare. Jayabalan et al. [44] suggested a scalable blockchain model that uses IPFS storage off the chain to protect healthcare data. They contend that current healthcare systems have problems maintaining data privacy and security because of the centralized storage of confidential patient information. Healthcare 4.0 represents the fourth industrial revolution in the healthcare industry, where the focus is on integrating digital technologies and data-driven solutions to revolutionize the delivery of healthcare services. This will enable a more personalized and precise approach to healthcare delivery, improving patient outcomes, enhancing patient experiences, and reducing the overall cost of care. Mahajan et al. [45] proposed incorporating blockchain technology and Healthcare 4.0 into cloud-based electronic health record (EHR) systems to enhance their security and accessibility. By integrating Healthcare 4.0 with blockchain technology, EHR systems can become more secure and efficient, enabling the safe and convenient exchange of patient information between various healthcare providers. Hathaliya et al. [46] emphasized the significance of safeguarding patient information in the Healthcare 4.0 era, characterized by the widespread use of mobile gadgets, wearable devices, and IoT. They suggest a method that entails biometric validation to permit access to health records solely to authorized staff. Remote patient monitoring (RPM) is a healthcare innovation that permits medical professionals to oversee and monitor a patient's health from a distance. RPM empowers healthcare providers to observe patients' well-being in real time, identify any fluctuations in their health status, and promptly intervene if needed. Subramanian et al. [47] proposed a novel approach to RPM that utilizes blockchain technology and the Internet of Things (IoT) to ensure secure and personalized care. They conducted a proof-of-concept study by developing a prototype system integrating Blockchain and IoT to monitor patients remotely. The study's results demonstrated the effectiveness and efficiency of their proposed system. Wadud et al. [48] added to the increasing research on using blockchain technology in the healthcare industry. They proposed a patient-centred system for remote patient monitoring that addresses concerns regarding privacy and ownership of patient data. Panwar et al. [49] pointed out the shortcomings of current methods. They stressed the importance of developing a stronger and more effective approach to handle the specific obstacles of protecting healthcare data in a Data Lake. To meet these needs, the paper introduces the BC-CCHS technique as a potential solution to guarantee data security while promoting efficient data exchange and cooperation between healthcare providers. Solution to improve the quality of remote patient monitoring in healthcare. Hoang et al. [50] described a system that utilizes smart contracts to facilitate the sharing of personal healthcare records (PHRs) securely and privately. Smart contracts ensure that the records are authentic, integral, and confidential. Additionally, the authors introduce a privacy-preserving method called "Proxy Re-Encryption," which allows authorized healthcare providers to access the PHRs without compromising the patient's identity. The primary objective of the paper by Bawany et al. [51] is to explore the potential of using blockchain technology to develop a secure and transparent platform for exchanging patient information, which can ultimately enhance healthcare outcomes. The proposed framework consists of three key elements: a patient portal, a healthcare provider portal, and a blockchain network. Vithanwattana et [52]. also suggested a security framework comprising three layers (physical, network, and application) to safeguard healthcare systems from security threats. They also emphasize the significance of training healthcare professionals and staff to adhere to security best practices and recognize security risks. Zheng et al. [53] suggested an approach which entails utilizing IoT gadgets to gather health-related information, which is then saved on a decentralized ledger like a blockchain. By leveraging DLTs, a secure and decentralized system is established to store and exchange health data, which may help resolve data security and confidentiality concerns. Moreover, the authors recommend utilizing smart contracts to automate agreements for sharing data between various healthcare providers, further simplifying the sharing process. Singh et al. [54] examined how blockchain technology can be used in healthcare. They recommended incorporating security and privacy measures to ensure the safe storage and sharing of patient medical records. The paper emphasizes the benefits of utilizing Blockchain in healthcare, including improved data security, compatibility, and patient confidentiality. The design by Uddin et al. [55] consists of several tiers: gathering information, analyzing data, making decisions, and executing actions. The agent that centres around the patient's needs manages the communication between these tiers and makes informed decisions based on patient data. Additionally, the suggested architecture employs a blockchain-driven system to guarantee the safety and confidentiality of patient information.

5 Proposed Methodology

In Fig. 4.4, the proposed healthcare architecture aims to streamline patient data management by using a unique healthcare ID. The government issues this ID and contains all pertinent health information, such as medical history, allergies, blood type, previous medications, chronic diseases, disabilities, and contact details. When patients visit a healthcare service provider, such as a hospital, they provide their healthcare ID to the attending doctor or nurse. This enables healthcare professionals to quickly and easily access the patient's health records, which is crucial in providing the best possible care. By having a centralized repository of patient data, healthcare providers can avoid potential errors or delays caused by incomplete or inaccurate patient information.

When a healthcare professional prescribes a new medication or adds new lab test records, as shown in Fig. 4.4, the information is securely sent to an application server. Afterwards, the server secures the fresh data by applying encryption and saves it onto the InterPlanetary File System (IPFS), an autonomous and dispersed system for storing files. The IPFS generates a content ID, which is then used to add the patient's information from their healthcare ID to a Solidity smart contract. This generates a transaction, which is validated by peer nodes and added to the private permissioned Blockchain. This secure and decentralized approach ensures that patient data is protected and can only be accessed by authorized healthcare professionals (Fig. 4.3).

Once the transaction containing the patient's health record is successfully added to the Blockchain shown in Fig. 4.5, the patient can access their health information through an easy-to-use application interface. This web interface is designed to connect both private and public healthcare service providers and practitioners. This connectivity can be essential in emergencies, where immediate access to the patient's medical information might mean the difference between life and death. This is especially true when healthcare professionals need access to formal healthcare interfaces to avoid treatment delays and subpar service. With this architecture,

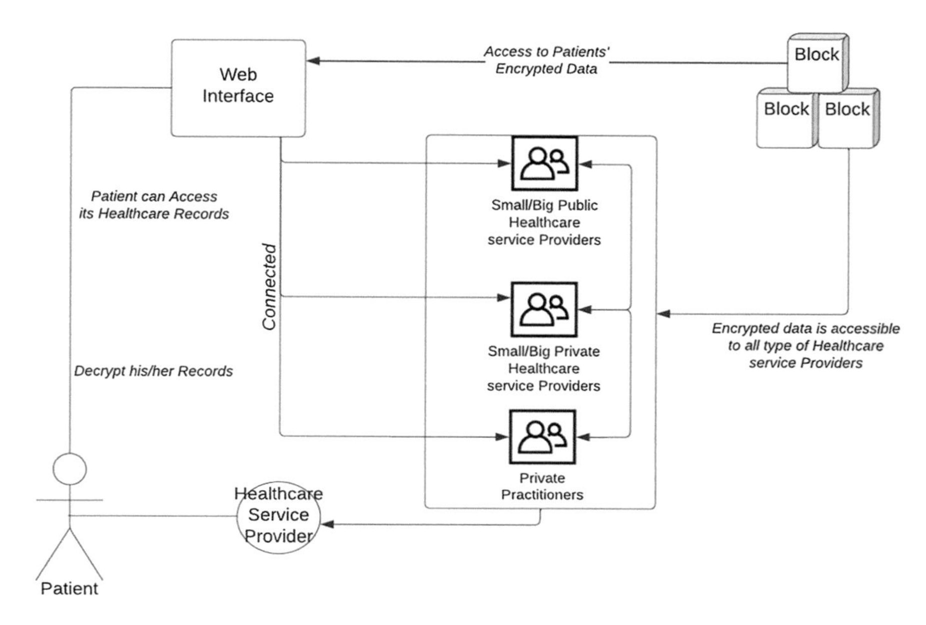

Fig. 4.3 Remote healthcare architecture

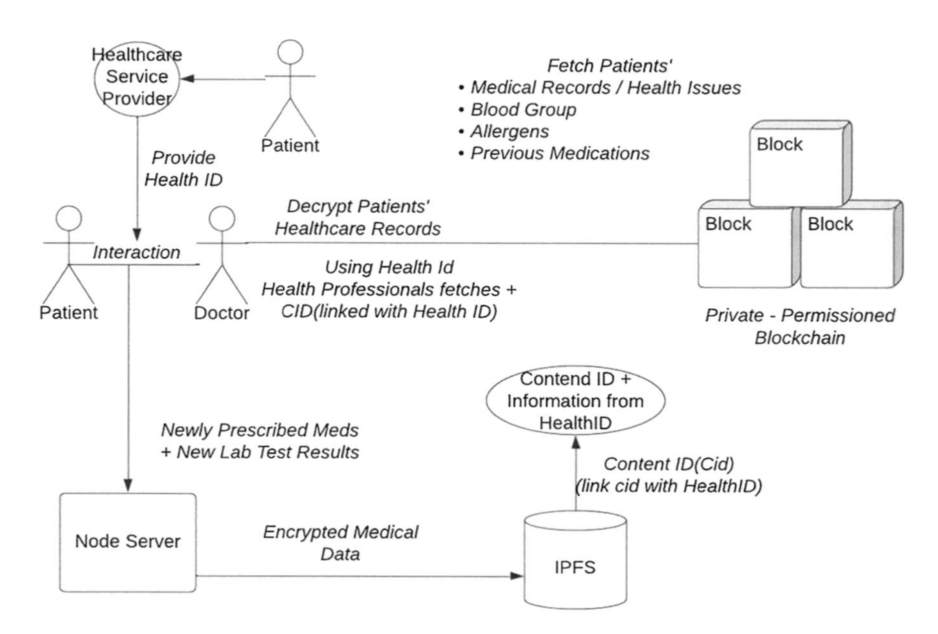

Fig. 4.4 Remote healthcare architecture

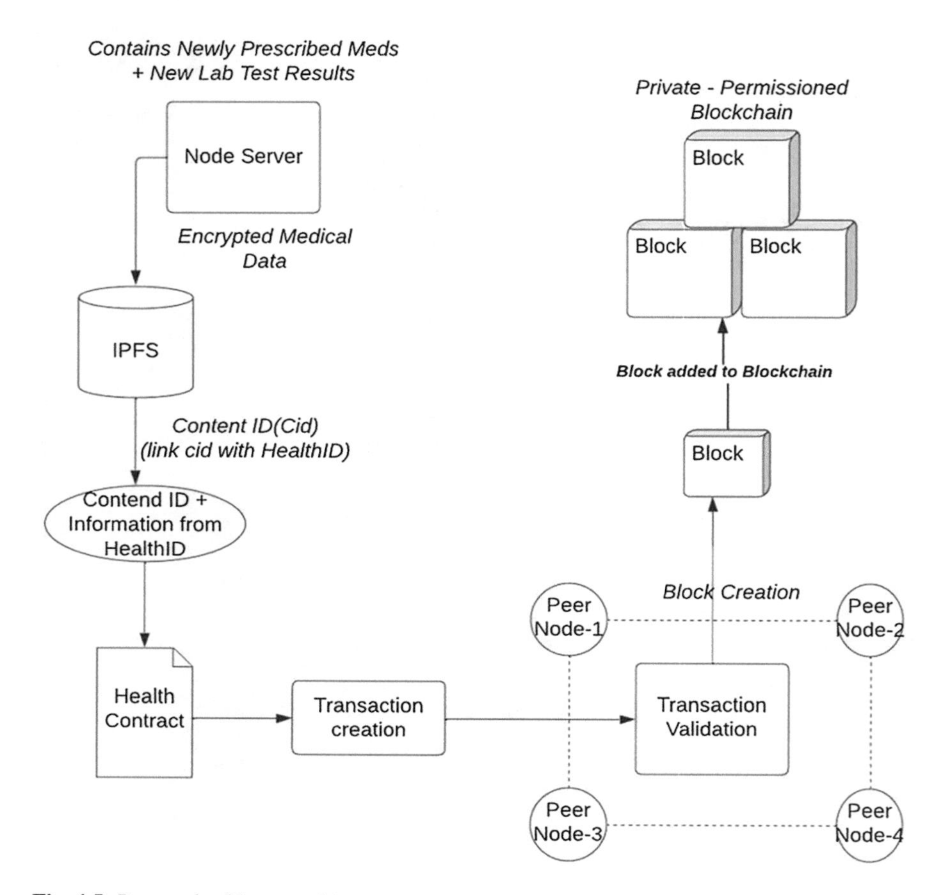

Fig. 4.5 Remote healthcare architecture

patients can rest assured that their medical data is accessible to authorized healthcare professionals regardless of where they seek treatment, as seen in Fig. 4.3.

Figure 4.3 shows that the proposed architecture for healthcare management offers numerous benefits to both patients and healthcare professionals. With a unique healthcare ID for each patient issued by the government, healthcare professionals can access all relevant health information quickly and easily. This includes medical history, allergies, blood type, previous medications, chronic diseases, disabilities, and contact information. This guarantees that medical professionals can obtain the necessary information for optimal healthcare, enhancing the standard of care and potentially rescuing lives during critical emergencies.

In addition, using a private permissioned blockchain shown in Fig. 4.3 connects various healthcare providers and practitioners, making it easier to access patient data in emergencies. This is especially important in cases where healthcare providers do not have access to structured healthcare interfaces, which can lead to delays in treatment. By securely storing patient information on the Blockchain, healthcare providers can access this information regardless of where the patient is receiving treatment, ensuring that patients receive the best possible care. Overall, the proposed architecture for healthcare management offers a secure, efficient, and effective way to manage patient health records, ensuring that patients receive the best possible care regardless of where they are receiving treatment which can be seen in Fig. 4.5. With a private permissioned blockchain and a unique healthcare ID for each patient, healthcare professionals can quickly and easily access patient information, improving health outcomes and potentially saving lives (Figs. 4.4 and 4.5).

The proposed architecture has several potential benefits, including:

- Improved patient care: By having a centralized database of patient health records accessible to all healthcare providers, doctors can make more informed decisions about patient care. They can see a patient's entire medical history, including any chronic conditions, allergies, and previous treatments, which can help them provide better care and avoid harmful drug interactions.
- Increased efficiency: With a centralized healthcare database, healthcare providers can quickly access patient information without waiting for records to be transferred between different providers. This can save time and resources and reduce the risk of errors.
- Improved data security: By using blockchain technology and IPFS to store patient data, the proposed architecture can provide high data security. Patient data may be safeguarded against illegal access and manipulation thanks to the decentralized structure of the Blockchain and the use of encryption.
- Increased patient autonomy: With access to their health records through a web interface, patients can take a more active role in managing their health. They can see their medical history, track their progress, and share their records with other healthcare providers as needed.
- Better communication between healthcare providers: The proposed architecture can improve communication and collaboration between providers by providing a centralized database of patient records that multiple healthcare providers can access. This can help ensure patients receive the best care, even when seeing multiple providers across different locations.

6 Benefits of Blockchain in Remote Healthcare

The integration of blockchain technology into remote healthcare has the potential to revolutionize the way healthcare services are delivered. Our proposed system utilizes a blockchain-based hyperledger implementation, ensuring secure storage of sensitive DICOM files. This approach offers multiple benefits for both patients and healthcare providers. By incorporating blockchain technology, we can bring about a transformative change in the healthcare industry.

- Enhanced data security: Using blockchain technology to store sensitive DICOM files dramatically improves data security by adding an extra layer of protection against unauthorized access and potential data breaches. This is accomplished by establishing a network with restricted access and privacy features, guaranteeing that only authorized users can reach the data. Consequently, implementing this system not only reinforces data security but also instils a greater sense of trust in the privacy of patient information.

- Increased data integrity: By utilizing blockchain technology to store DICOM files, the system can guarantee increased data integrity by implementing tamper-proof and immutable storage mechanisms. This ensures that the data remains unchanged and secure throughout its storage, and any attempts to alter or manipulate the information will be immediately detected. Furthermore, using Hyperledger Fabric/Hyperledger Besu adds another layer of assurance to the data's integrity, as authorized parties verify and validate the data, further strengthening the trustworthiness and reliability of the system. The use of advanced technologies to maintain the accuracy and security of data is remarkable progress in the healthcare industry. It can completely transform how sensitive patient information is stored and protected.
- Improved transparency: Implementing a blockchain system in managing DICOM files can offer improved transparency and visibility into the access and utilization of sensitive patient data. This transparency is crucial in ensuring regulatory compliance and accuracy, vital in the healthcare industry where legal and ethical standards must be upheld. The blockchain-based system provides an immutable and tamper-proof record of every access and modification made to DICOM files, enabling auditors to verify the legitimacy and accuracy of the data. Blockchain technology in healthcare can enhance trust and responsibility among healthcare providers by providing a transparent and visible system. This allows patients to have confidence that their confidential information is being managed securely and ethically. The increased transparency brought by blockchain technology is a valuable benefit, ensuring adherence to regulations and cultivating trust between patients and healthcare providers.
- Reduced costs: Implementing a blockchain-powered platform to store and handle DICOM files could notably impact decreasing expenses linked to transactions, storage, and administrative duties. Utilizing zero-fee transactions reduces the costs of accessing and storing the files, resulting in a more economical approach for healthcare providers and patients. Moreover, utilizing a blockchain-based system can also help reduce administrative expenses related to data management, as the system is automated and requires less human intervention. This translates into a more streamlined and efficient process for managing healthcare data, ultimately leading to lower costs for healthcare providers and patients. The potential for cost reduction is a significant benefit of implementing such innovative and advanced technology in the healthcare industry.
- Improved efficiency: Incorporating automation and continuous delivery/integration into the system for managing DICOM files using blockchain technology can enhance data management and processing efficiency. This, in turn, can alleviate healthcare providers' workload and enable quicker data access. The automated nature of the system ensures that data is managed and processed without the need

for extensive manual intervention, making the entire process more streamlined and efficient. The continuous delivery/integration approach further enhances the system's efficiency by enabling frequent and rapid updates, ensuring that it remains up-to-date and capable of handling the latest data management requirements. The resulting efficiency improvements are crucial in today's fast-paced healthcare industry, where time is often of the essence, and prompt access to data can make all the difference in providing adequate care to patients.

- Improved user experience: Using a user interface, the system for managing DICOM files through blockchain technology can significantly enhance the user experience, making it more accessible and user-friendly for healthcare providers and patients. The interface simplifies the process of interacting with and accessing DICOM files, ensuring that the data can be easily accessed and managed. The streamlined user experience can increase data management and processing efficiency, as users can quickly and efficiently navigate through the system's various features and functionalities. This results in a more intuitive and engaging user experience, promoting greater engagement with the system and its benefits. Ultimately, incorporating a user interface in the blockchain-based system for managing DICOM files can revolutionize how healthcare providers and patients interact with and manage sensitive patient information, providing a more seamless and engaging user experience.
- Enhanced scalability: Incorporating a blockchain-based system for managing DICOM files can significantly enhance the scalability of data storage and processing, a critical requirement in remote healthcare settings. Utilizing blockchain technology allows the system to manage substantial amounts of data effectively, offering an exceedingly adaptable approach for storing and retrieving confidential patient data. This is especially important in remote healthcare, where efficient and scalable data management is paramount. The blockchain-based system's ability to handle increasing amounts of data without compromising its performance ensures that it remains reliable and effective, even as data volumes grow. The enhanced scalability of the system can help healthcare providers and patients access and manage data more efficiently, ultimately improving the quality of care delivered to patients. Therefore, the incorporation of blockchain technology represents a significant breakthrough in remote healthcare, offering a highly scalable and efficient solution for managing sensitive patient information.

7 Future Challenges of the Topic

Although blockchain technology holds significant promise in transforming the healthcare industry, several hurdles must be overcome to adopt it widely in remote healthcare. Realizing this technology's potential depends on overcoming various future challenges associated with its implementation. Various obstacles, including technical and regulatory concerns, can hinder the smooth incorporation of blockchain technology into remote healthcare environments. Overcoming these challenges is essential to facilitate the successful implementation of blockchain technology in remote healthcare settings. These challenges include establishing a robust technical infrastructure capable of supporting blockchain systems, compliance with existing regulations, ensuring data privacy and security, and addressing interoperability issues between different healthcare systems. By overcoming these obstacles, blockchain technology will be successfully incorporated into remote healthcare, improving the standard of care given to patients in these situations.

- Integration with existing systems: One of the significant hurdles to adopting blockchain technology in remote healthcare is integrating this innovative technology with the existing healthcare systems. Many healthcare providers depend on outdated systems that may need to work better with blockchain technology. This lack of compatibility creates difficulties when trying to incorporate Blockchain into the current infrastructure, which means there is a need for substantial investment in new systems and infrastructure. Integrating blockchain technology with these outdated systems can be a complicated, time-consuming, and expensive process that might even require a complete revamp of the existing healthcare infrastructure. Therefore, healthcare providers must invest in upgrading their infrastructure to support blockchain technology, ensuring seamless integration and maximizing the technology's potential benefits in remote healthcare settings. Only by addressing this challenge can healthcare providers take full advantage of blockchain technology's transformative potential in remote healthcare.
- Regulatory and legal barriers: Introducing blockchain technology in the healthcare sector brings up various regulatory and legal obstacles that must be resolved to guarantee its secure and efficient application in remote healthcare. This technology gives rise to multiple worries, such as safeguarding data privacy and security, determining liability in the event of data breaches, and adhering to current healthcare regulations. Addressing these challenges is critical to ensure healthcare providers can utilize blockchain technology safely and effectively in remote healthcare settings. The regulatory and legal barriers surrounding the implementation of blockchain technology require a multidisciplinary approach involving various stakeholders, including healthcare providers, policy-makers, legal experts, and regulatory bodies. This approach will ensure blockchain technology is implemented within existing regulations and laws, protecting patients' data privacy and security. By addressing these regulatory and legal issues, healthcare providers can take full advantage of the transformative potential of blockchain technology, enhancing the quality of care delivered in remote healthcare settings.
- Technical expertise: Implementing blockchain technology in remote healthcare requires significant technical expertise. This expertise is necessary for developing and maintaining the blockchain infrastructure and training healthcare providers on how to use the technology effectively. The lack of technical expertise poses a major obstacle to adopting blockchain technology, particularly for smaller healthcare providers with limited resources. Developing and maintaining

the necessary blockchain infrastructure can be complicated and demands specialized knowledge and skills. To overcome this challenge, healthcare providers must allocate resources to enhance their technical capabilities and acquire the necessary skills.

Moreover, training healthcare providers to utilize the technology effectively requires significant time and resources. This training must be continuous to ensure that healthcare providers stay updated with the latest advancements in blockchain technology. By addressing this challenge, healthcare providers can fully unlock the potential benefits of blockchain technology in remote healthcare, thereby enhancing the quality of care provided to patients.

- Interoperability: Implementing blockchain technology in remote healthcare faces a notable obstacle in interoperability. For blockchain technology to be effective, it must be able to interoperate with other healthcare systems and data sources. However, this cannot be easy because of the diversity of healthcare systems and data sources currently in use. Healthcare providers must ensure their blockchain infrastructure can interoperate with existing systems to avoid data duplication and other inefficiencies. Additionally, the need for more standardization in the healthcare industry can make it challenging to ensure blockchain technology can interoperate with different systems and data sources. Addressing these interoperability challenges will require collaboration between healthcare providers, technology companies, and regulatory agencies. By collaborating and working collectively, we can only achieve a healthcare system that effectively utilizes blockchain technology in remote healthcare.
- Cost: The expense of introducing blockchain technology in distant healthcare can pose a notable obstacle to its adoption, particularly for smaller healthcare providers who have restricted budgets. While blockchain technology can decrease costs in the long term, the initial investment needed to embrace the technology can be substantial. This encompasses the expenses associated with creating and sustaining the blockchain system and the costs of educating healthcare providers to utilize the technology proficiently. Also, smaller healthcare organizations might need more funds to invest in the technical know-how and infrastructure needed to use blockchain technology. Therefore, finding ways to reduce the costs associated with implementing blockchain technology in remote healthcare will be critical to its widespread adoption.

Although blockchain technology holds significant promise for revolutionizing remote healthcare, its widespread adoption faces several challenges. Addressing these challenges will require a collaborative effort among various stakeholders, including healthcare providers, technology companies, and regulatory bodies. Addressing technical, regulatory, legal, interoperability, and cost issues will be necessary to enable blockchain technology's safe and effective adoption in remote healthcare. Nevertheless, given the potential benefits that blockchain technology can offer in remote healthcare, overcoming these challenges is worth the effort. By working together, stakeholders can leverage the strengths of blockchain technology to transform remote healthcare and improve patient outcomes.

8 Conclusion

In conclusion, The proposed solution for storing sensitive DICOM files in a Hyperledger blockchain system combined with Grafana Loki monitoring, Hyperledger Fabric/Hyperledger Besu, vulnerability checks, metadata generation, permission and privacy-enabled network, a user interface, zero-fee transactions, and continuous delivery/integration, provides numerous advantages for managing remote healthcare data. Utilizing blockchain technology can enhance the security and reliability of data, provide clarity regarding data access and utilization, reduce expenses, boost effectiveness and scalability, and enhance overall user satisfaction. Despite some challenges, such as regulatory and legal barriers and the need for technical expertise, this system has great potential for improving remote healthcare data management. A collaborative effort between healthcare providers, technology companies, and regulatory bodies is necessary to implement this solution and maximize its benefits.

References

1. Casino, F., Dasaklis, T. K., & Patsakis, C. (2019). A systematic literature review of blockchain-based applications: Current status, classification and open issues. *Telematics and Informatics, 36*, 55–81. https://doi.org/10.1016/j.tele.2018.11.006
2. Zheng, Z., Xie, S., Dai, H.-N., Chen, X., & Wang, H. (2017). *An overview of blockchain technology: Architecture, consensus, and future trends*. https://doi.org/10.1109/BigDataCongress.2017.85
3. Zarrin, J., Wen Phang, H., Babu Saheer, L., & Zarrin, B. (2021). Blockchain for decentralization of internet: Prospects, trends, and challenges. *Cluster Computing, 24*, 2841. https://doi.org/10.1007/s10586-021-03301-8
4. Hoffman, M. R., Ibáñez, L.-D., & Simperl, E. (2020). Toward a formal scholarly understanding of blockchain-mediated decentralization: A systematic review and a framework. *Frontiers in Blockchain, 3*. https://doi.org/10.3389/fbloc.2020.00035
5. Shrimali, B., & Patel, H. B. (2022). Blockchain state-of-the-art: Architecture, use cases, consensus, challenges and opportunities. *Journal of King Saud University – Computer and Information Sciences, 34*(9), 6793–6807. https://doi.org/10.1016/j.jksuci.2021.08.005
6. Alharby, M., & Moorsel, A. v. (2017). *Blockchain based smart contracts: A systematic mapping study*. Computer Science Information Technology (CS IT). https://doi.org/10.5121/csit.2017.71011
7. Khan, S. N., Loukil, F., Ghedira-Guegan, C., Benkhelifa, E., & Bani-Hani, A. (2021). Blockchain smart contracts: Applications, challenges, and future trends. *Peer-to-Peer Networking and Applications, 14*, 2901. https://doi.org/10.1007/s12083-021-01127-0
8. Newaz, A., Sikder, A., Rahman, M., & Uluagac, A. (2021). A survey on security and privacy issues in modern healthcare systems: Attacks and defenses. *ACM Transactions on Computing for Healthcare, 2*(3), 1–44.
9. Chaudhry, N., & Yousaf, M. M. (2018). Consensus algorithms in blockchain: Comparative analysis, challenges and opportunities. In *2018 12th international conference on open source systems and technologies (ICOSST)*. https://doi.org/10.1109/icosst.2018.8632190

10. Mingxiao, D., Xiaofeng, M., Zhe, Z., Xiangwei, W., & Qijun, C. (2017). A review on consensus algorithm of blockchain. In *2017 IEEE international conference on systems, man, and cybernetics (SMC)*. https://doi.org/10.1109/smc.2017.8123011
11. Nguyen, T., & Kim, K. (2018). A survey about consensus algorithms used in blockchain. *Journal of Information Processing Systems, 14*, 101–128. https://doi.org/10.3745/JIPS.01.0024
12. Archip, A., Botezatu, N., Şerban, E., Herghelegiu, P. C., & Zală, A. (2016). An IoT based system for remote patient monitoring. In *17th international carpathian control conference (ICCC)* (pp. 1–6).
13. Sheikh, H., Azmathullah, R., & Rizwan, F. (2018). Proof-of-work vs proof-of-stake: A comparative analysis and an approach to blockchain consensus mechanism. *International Journal for Research in Applied Science Engineering Technology (IJRASET), 887*, 2321–9653.
14. Gemeliarana, I., & Sari, R. (2019). *Evaluation of proof of work (pow) blockchains security network on selfish mining*. https://doi.org/10.1109/ISRITI.2018.8864381
15. Gemeliarana, I. G. A. K., & Sari, R. F. (2018). Evaluation of proof of work (pow) blockchains security network on selfish mining. In *2018 international seminar on research of information technology and intelligent systems (IS-RITI)*. https://doi.org/10.1109/isriti.2018.8864381
16. Nguyen, C., Dinh Thai, H., Nguyen, D., Niyato, D., Nguyen, H., & Dutkiewicz, E. (2019). Proof-of-stake consensus mechanisms for future blockchain networks: Fundamentals, applications and opportunities. *IEEE Access, 7*, 1–1. https://doi.org/10.1109/ACCESS.2019.2925010
17. Ray, P. P., Dash, D., Salah, K., & Kumar, N. (2021). Blockchain for IoTbased healthcare: Background, consensus, platforms, and use cases. *IEEE Systems Journal, 15*, 85–94.
18. Gai, K., Qiu, M., Xiong, Z., & Liu, M. (2018). Privacy-preserving multichannel communication in edge-of-things. *Future Generation Computer Systems, 85*, 190–200.
19. Hu, Q., Yan, B., Han, Y., & Yu, J. (2021). An improved delegated proof of stake consensus algorithm. *Procedia Computer Science, 187*, 341–346. https://doi.org/10.1016/j.procs.2021.04.109
20. Moin, S., Karim, A., Safdar, Z., Safdar, K., Ahmed, E., & Imran, M. (2019). Securing iots in distributed blockchain: Analysis, requirements and open issues. *Future Generation Computer Systems*, 325–343.
21. Chen, Y., & Liu, F. (2022). Research on improvement of dpos consensus mechanism in collaborative governance of network public opinion. *Peer-to-Peer Networking and Applications, 15*, 1849–1861. https://doi.org/10.1007/s12083-022-01320-9
22. Castro, M., & Liskov, B. (2002). Practical byzantine fault tolerance and proactive recovery. *ACM Transactions on Computer Systems, 20*, 398–461. https://doi.org/10.1145/571637.571640
23. Tandon, A., Dhir, A., Islam, A. N., & Mäntymäki, M. (2020). Blockchain in healthcare: A systematic literature review, synthesizing framework and future research agenda. *Computers in Industry, 122*, 103290.
24. Borda, M., Grishchenko, N., & Kowalczyk-Rólczyńska, P. (2022). Patient readiness for remote healthcare services in the context of the covid-19 pandemic: Evidence from european countries. *Frontiers in Public Health, 10*. https://doi.org/10.3389/fpubh.2022.846641
25. Smolić, Š., Blaževski, N., & Fabijančić, M. (2022). Remote healthcare during the covid-19 pandemic: Findings for older adults in 27 European countries and Israel. *Frontiers in Public Health, 10*. https://doi.org/10.3389/fpubh.2022.921379
26. DePuccio, M. J., Gaughan, A. A., Shiu-Yee, K., & McAlearney, A. S. (2022). Doctoring from home: Physicians' perspectives on the advantages of remote care delivery during the covid-19 pandemic. *PLoS One, 17*, e0269264. https://doi.org/10.1371/journal.pone.0269264
27. Haleem, A., Javaid, M., Singh, R. P., & Suman, R. (2021). Telemedicine for healthcare: Capabilities, features, barriers, and applications. *Sensors International, 2*, 100117. https://doi.org/10.1016/j.sintl.2021.100117
28. Al-Ahmadi, B. (2019). Blockchain based remote patient monitoring system. *Journal of King Abdulaziz University Computing and Information Technology Sciences, 8*, 111–118. https://doi.org/10.4197/Comp.8-2.8

29. Tandon, A., Dhir, A., Islam, N., & Mäntymäki, M. (2020). Blockchain in healthcare: A systematic literature review, synthesizing framework and future research agenda. *Computers in Industry, 122*, 103290. https://doi.org/10.1016/j.compind.2020.103290
30. Khezr, S., Moniruzzaman, M., Yassine, A., & Benlamri, R. (2019). Blockchain technology in healthcare: A comprehensive review and directions for future research. *Applied Sciences, 9*, 1736. https://doi.org/10.3390/app9091736
31. Wu, H., Dwivedi, A. D., & Srivastava, G. (2021). Security and privacy of patient information in medical systems based on blockchain technology. *ACM Transactions on Multimedia Computing, Communications, and Applications, 17*, 1–17.
32. Wang, M., Guo, Y., Zhang, C., Wang, C., Huang, H., & Jia, X. (2021). *MedShare: A privacy-preserving medical data sharing system by using Blockchain*. IEEE Transactions on Services Computing.
33. Ellouze, F., Fersi, G., & Jmaiel, M. (2020). Blockchain for internet of medical things: A technical review. In M. Jmaiel, M. Mokhtari, B. Abdulrazak, H. Aloulou, & S. Kallel (Eds.), *Proceedings of the impact of digital technologies on public health in developed and developing countries* (pp. 259–267). Springer.
34. Ben Fekih, R., & Lahami, M. (2020). *Application of blockchain technology in healthcare: A comprehensive study* (pp. 268–276). Lecture Notes in Computer Science. https://doi.org/10.1007/978-3-030-51517-123
35. Cornelius, K. (2021). Betraying blockchain: Accountability, transparency and document standards for non-fungible tokens (nfts). *Information, 12*, 358. https://doi.org/10.3390/info12090358
36. Farnaghi, M., & Mansourian, A. (2020). Blockchain, an enabling technology for transparent and accountable decentralized public participatory gis. *Cities, 105*, 102850. https://doi.org/10.1016/j.cities.2020.102850
37. Rather, I. H., & Idrees, S. M. (2021). Blockchain technology and its applications in the healthcare sector. In *Blockchain for healthcare systems* (pp. 17–25). CRC Press.
38. Batubara, F., Ubacht, J., & Janssen, M. (2019). *Unraveling transparency and accountability in blockchain* (pp. 204–213). https://doi.org/10.1145/3325112.3325262
39. Rahman, M. A., Hossain, M. S., Islam, M. S., Alrajeh, N. A., & Muhammad, G. (2020). Secure and provenance enhanced internet of health things framework: A blockchain managed federated learning approach. *IEEE Access, 8*, 205071–205087.
40. Peregrin, T. (2021). Managing hipaa compliance includes legal and ethical considerations. *Journal of the Academy of Nutrition and Dietetics, 121*, 327–329. https://doi.org/10.1016/j.jand.2020.11.012
41. Chhikara, D., Rana, S., Mishra, A., & Mishra, D. (2022). Blockchain-driven authorized data access mechanism for digital healthcare. *Journal of Systems Architecture, 131*, 102714. https://doi.org/10.1016/j.sysarc.2022.102714
42. Hathaliya, J., Sharma, P., Tanwar, S., & Gupta, R. (2019). Blockchain-based remote patient monitoring in healthcare 4.0. In *2019 IEEE 9th international conference on advanced computing (IACC)*. https://doi.org/10.1109/iacc48062.2019.8971593
43. Sarier, N. D. (2022). Privacy preserving biometric authentication on the blockchain for smart healthcare. *Pervasive and Mobile Computing*, 101683. https://doi.org/10.1016/j.pmcj.2022.101683
44. Jayabalan, J., & Jeyanthi, N. (2022). Scalable blockchain model using off-chain ipfs storage for healthcare data security and privacy. *Journal of Parallel and Distributed Computing, 164*, 152. https://doi.org/10.1016/j.jpdc.2022.03.009
45. Mahajan, H. B., Rashid, A. S., Junnarkar, A. A., Uke, N., Deshpande, S. D., Futane, P. R., Alkhayyat, A., & Alhayani, B. (2022). Integration of healthcare 4.0 and blockchain into secure cloud-based electronic health records systems. *Applied Nanoscience, 13*, 2329. https://doi.org/10.1007/s13204-021-02164-0

46. Hathaliya, J. J., Tanwar, S., Tyagi, S., & Kumar, N. (2019). Securing electronics healthcare records in healthcare 4.0: A biometric-based approach. *Computers Electrical Engineering, 76*, 398–410. https://doi.org/10.1016/j.compeleceng.2019.04.017
47. Subramanian, G., & Thampy, A. S. (2021). Implementation of blockchain consortium to prioritize diabetes patients' healthcare in pandemic situations. *IEEE Access, 9*, 162459–162475. https://doi.org/10.1109/access.2021.3132302
48. Wadud, M. A. H., Amir-Ul-Haque Bhuiyan, T. M., Uddin, M. A., & Rahman, M. M. (2020). *A patient centric agent assisted private blockchain on hyperledger fabric for managing remote patient monitoring*. https://doi.org/10.1109/ICECE51571.2020.9393124
49. Panwar, A., & Bhatnagar, V. (2021). A cognitive approach for blockchain-based cryptographic curve hash signature (bc-cchs) technique to secure healthcare data in data lake. *Soft Computing*, 1. https://doi.org/10.1007/s00500-021-06513-7
50. Hoang, H. D., Hien, D. T. T., Nhut, T. C., Quyen, P. D. T., Duy, P. T., & Pham, V.-H. (2021). *A blockchain-based secured and privacy-preserved personal healthcare record exchange system*. https://doi.org/10.1109/ICMLANT53170.2021.9690542
51. Bawany, N. Z., Qamar, T., Tariq, H., & Adnan, S. (2022). Integrating healthcare services using blockchain-based telehealth framework. *IEEE Access, 10*, 1–1. https://doi.org/10.1109/access.2022.3161944
52. Vithanwattana, N., Karthick, G., Mapp, G., & George, C. (2021). *Exploring a new security framework for future healthcare systems*. https://doi.org/10.1109/GCWkshps52748.2021.9681967
53. Zheng, X., Sun, S., Mukkamala, R. R., Vatrapu, R., & Ordieres-Meré, J. (2019). Accelerating health data sharing: A solution based on the internet of things and distributed ledger technologies. *Journal of Medical Internet Research, 21*, e13583. https://doi.org/10.2196/13583
54. Singh, J., & Ghai, K. (2021). *Security and privacy mechanisms for the new generation healthcare applications using blockchain technology*. https://doi.org/10.1109/ICRITO51393.2021.9596107
55. Uddin, M. A., Stranieri, A., Gondal, I., & Balasubramanian, V. (2018). Continuous patient monitoring with a patient centric agent: A block architecture. *IEEE Access, 6*, 32700–32726. https://doi.org/10.1109/access.2018.2846779

Chapter 5
Smart Contract Vulnerabilities: Exploring the Technical and Economic Aspects

Deepak Dhillon, Diksha, and Deepti Mehrotra

1 Introduction

In the rapidly evolving landscape of blockchain technology, smart contracts have emerged as a pivotal innovation, poised to redefine the fabric of digital agreements. These self-executing contracts encapsulate the convergence of technical intricacies and economic implications, offering a new paradigm for automating, verifying, and executing transactions [1]. This chapter embarks on a comprehensive exploration of the multifaceted realm of smart contracts, delving into their inherent significance, the intricate interplay of technical and economic dimensions, and a holistic perspective on their vulnerabilities and limitations.

At the core of this exploration lies the profound significance of smart contracts. These digitally encoded agreements not only streamline traditional contractual processes but also establish a trustless and tamper-proof environment for executing transactions. By seamlessly automating processes and reducing the reliance on intermediaries, smart contracts hold the potential to revolutionize sectors ranging from finance to logistics. The introduction of smart contracts transcends technical prowess; it intertwines with economic implications. The automation, transparency, and efficiency they offer have profound economic ramifications, potentially disrupting industries, and reshaping business operations. This convergence of technology and economics warrants a comprehensive analysis that encompasses both domains.

D. Dhillon (✉)
Amity Institute of Defence Technology, Noida, UP, India
e-mail: deepak10@s.amity.edu

Diksha
Banasthali Vidyapeeth, Jaipur, Rajasthan, India

D. Mehrotra
Amity School of Engineering and Technology, Noida, UP, India

S. M. Idrees, M. Nowostawski (eds.), *Blockchain Transformations*, Signals and Communication Technology, https://doi.org/10.1007/978-3-031-49593-9_5

While the promise of smart contracts is enticing, their implementation introduces vulnerabilities that demand meticulous scrutiny. This chapter takes a holistic approach, unveiling the vulnerabilities within smart contracts that could compromise their integrity and security. Additionally, it navigates the practical limitations and challenges faced across various industries, providing a balanced perspective on their transformative potential.

As we embark on this journey, the intricate tapestry of smart contracts unfolds, revealing their potential to revolutionize conventional business agreements. By delving into their significance, dissecting their technical and economic dimensions, and acknowledging their vulnerabilities and limitations, this chapter strives to foster a comprehensive understanding that will guide informed decisions in harnessing the transformative power of smart contracts.

2 Economic Aspects of Smart Contracts

Smart contracts offer numerous advantages over traditional contracts, including efficiency, cost savings, transparency, immutability, increased trust, automation, security, and disintermediation. They automate contract execution, minimizing reliance on middlemen and saving time and costs. Smart contracts are transparent and immutable, providing real-time visibility into contract terms and ensuring the reliability of the agreement. The use of blockchain technology enhances trust among contract participants by eliminating the need for a central authority [11] and providing equal access and verification rights [7]. Automation simplifies business processes and integrates with other systems. Smart contracts benefit from the security of blockchain technology, ensuring the integrity and reliability of the agreement [3]. They also enable disintermediation, reducing costs and speeding up contract execution.

Smart contracts have potential use cases in various industries. In finance, they can streamline transactions, automate insurance claims processing, and facilitate peer-to-peer lending. In the real estate industry, smart contracts can automatically transfer property ownership, simplify lease agreements, and minimize the need for intermediaries. Supply chain management can benefit from smart contracts through automated track and trace of goods and streamlined payment processing.

However, challenges exist, such as the need for standardization across industries and the development of legal and regulatory frameworks to support smart contract usage [2]. Addressing these obstacles will be pivotal for the widespread acceptance and realization of the complete potential of smart contracts in various sectors.

3 Limitations and Challenges of Implementing Smart Contracts

Smart contracts in the blockchain ecosystem have economic aspects that offer advantages over traditional contracts [3], but they also come with limitations and challenges. They reduce transaction costs and enhance efficiency by automating contract execution, minimizing the reliance on intermediaries and third-party validation. This saves time and money while increasing transparency and reducing fraud. However, implementing smart contracts can be complex due to the requirement of blockchain technology [4]. Blockchain is still in its early stages, and debates on the best design and implementation approaches are ongoing. In addition, the legal and regulatory structures for smart contracts are developing, causing ambiguity for businesses and users.

Another limitation is the potential for errors or bugs in the code [6], which can have severe consequences due to the self-executing nature of smart contracts. Auditing and testing are crucial to ensure error-free operation. Scalability is another challenge, as an increasing number of users can cause delays and higher transaction fees. Scaling solutions like sharding and sidechains are being explored to address this challenge.

Adopting smart contracts requires a shift in mindset and culture. Traditional contracts rely on legal and financial expertise, while smart contracts require programming and technical skills [7]. Education and training are essential to aid users in comprehending the advantages, restrictions, and optimal utilization of smart contracts.

4 Applications of Smart Contracts Across Industries (Fig. 5.1)

Smart contracts have propelled a wave of innovation across diverse industries, promising automation, efficiency, and transparency [13]. However, each sector also faces its unique set of challenges and limitations in fully embracing this transformative technology.

1. Food Industry: Supply Chain Management and Integration Challenges

The food industry's intricate supply chains could benefit from smart contract automation, enhancing traceability and reducing fraud. Yet, challenges lie in integrating various stakeholders onto blockchain platforms and navigating complexities in data sharing.

2. Real Estate: Legal Hurdles, Standardization, and Adoption

Real estate transactions can leverage smart contracts to streamline property transactions, reducing intermediary involvement. However, legal and regulatory

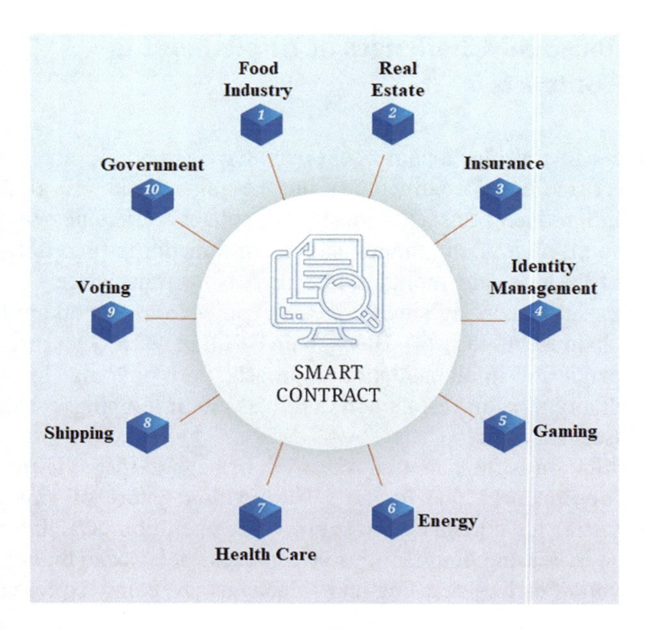

Fig. 5.1 Applications of smart contracts across industries

hurdles, along with the need for standardized processes, impede widespread adoption.

3. Insurance: Integration Complexities and Regulatory Challenges

Smart contracts can expedite claims processing and policy management in the insurance sector [14]. Yet, integrating these contracts with existing systems and regulations poses significant challenges, hampering their full implementation.

4. Identity Management: Adoption and Interoperability Dilemmas

Smart contracts hold potential for secure and decentralized identity management. However, achieving widespread adoption and seamless integration with existing identity systems remains a complex endeavour.

5. Gaming: Limited Adoption and Regulatory Constraints

In the gaming industry, smart contracts can offer transparency and fairness in transactions and rewards. Yet, limited adoption by gaming companies, coupled with evolving regulatory constraints, hampers their extensive implementation.

6. Energy: Industry Adoption and Integration Hurdles

Smart contracts can revolutionize energy trading and distribution by automating transactions and ensuring accountability. Nevertheless, industry-wide adoption and integration with legacy systems pose significant challenges.

7. Healthcare: Challenges in Adoption and Integration

The healthcare sector could benefit from smart contracts for secure patient data sharing and drug traceability. However, adopting these contracts within the complex healthcare ecosystem and integrating them with existing systems present formidable hurdles.

8. Shipping: Industry Acceptance and Regulatory Challenges

Smart contracts hold potential for streamlining shipping and logistics operations. Industry-wide acceptance and navigating regulatory intricacies, however, remain significant roadblocks.

9. Voting: Adoption, Integration, and Regulatory Hurdles

Smart contracts can enhance transparency and security in voting systems. Yet, their adoption hinges on overcoming challenges related to voter acceptance, integration with existing systems, and regulatory concerns.

10. Government: Citizen Engagement, Integration, and Regulatory Landscape

Smart contracts could transform government operations, from digital identities to automated public services. Yet, citizen acceptance, integration complexities, and evolving regulatory frameworks pose substantial challenges.

In each sector, the potential benefits of smart contracts are undeniable. However, the path to their widespread adoption is riddled with challenges that demand collaboration among technology developers, industry stakeholders, and regulatory bodies. As smart contracts continue to evolve, finding solutions to these challenges is essential in unlocking their transformative potential across industries.

4.1 Potential Benefits and Challenges of Smart Contracts in Various Industries

A comparative overview of how smart contracts can revolutionize diverse industries, juxtaposed with the hurdles impeding their comprehensive adoption (Table 5.1).

An exploration of how smart contracts hold transformative potential across industries, countered by challenges necessitating collaborative solutions among stakeholders and regulatory bodies.

Table 5.1 Potential benefits and challenges of smart contracts in various industries

Industry	Potential benefits of smart contracts	Challenges and limitations
Food	Traceability, reduced fraud	Integration, data sharing complexity
Real estate	Streamlined transactions, reduced intermediaries	Legal hurdles, lack of standardization
Insurance	Claims processing, policy management	Integration, regulatory compliance
Identity management	Secure & decentralized identity	Adoption, interoperability
Gaming	Transparency, fairness	Limited adoption, regulatory constraints
Energy	Automated trading, accountability	Industry adoption, legacy system integration
Healthcare	Patient data sharing, drug traceability	Adoption, integration within healthcare ecosystem
Shipping	Streamlined operations	Industry acceptance, regulatory challenges
Voting	Transparency, security	Adoption, integration, regulatory concerns
Government	Digital identities, automated services	Citizen engagement, integration, regulatory landscape

4.2 *High-Risk Factors and Mitigation Strategies Across Industries*

A comprehensive overview of high-risk factors inherent in various industries, accompanied by potential mitigation strategies to address these challenges effectively (Table 5.2).

A proactive examination of high-risk factors across industries, coupled with strategic approaches to mitigate challenges and enhance the successful integration of smart contracts while ensuring industry growth and security.

5 Vulnerabilities Within Smart Contracts

Smart contracts, hailed for their potential to streamline processes and foster transparency, are not exempt from vulnerabilities that could jeopardize their integrity and security. This section delves into several significant vulnerabilities that can materialize within smart contracts, magnifying the necessity for robust security measures and best practices in their development [8].

On 17 June 2016, a notable example occurred when a "Blackhat" hacker exploited a vulnerability in The DAO's smart contract, siphoning $150 million worth of ETH [9]. This incident underscored the complexity of ensuring secure

Table 5.2 High-risk factors and mitigation strategies across industries

Industry	High-risk factors	Mitigation strategies
Food	Supply chain disruptions, counterfeit products	Blockchain-based traceability, supplier verification
Real estate	Property title disputes, fraudulent transactions	Blockchain-based property records, legal reforms
Insurance	Mismanagement of claims, regulatory non-compliance	Automated claims processing, regulatory compliance
Identity management	Data breaches, lack of standardized protocols	Decentralized identity solutions, encryption
Gaming	Security breaches, lack of player protection	Enhanced cybersecurity measures, fair play audits
Energy	Market volatility, cybersecurity vulnerabilities	Smart contract-based energy trading, robust IT security
Healthcare	Patient data breaches, interoperability issues	Encrypted patient data storage, standardized protocols
Shipping	Cargo theft, logistical breakdowns	IoT-enabled cargo tracking, contingency planning
Voting	Cyberattacks, voter manipulation	Blockchain-based secure voting, multi-factor authentication
Government	Data privacy breaches, digital divide	Strong data protection policies, digital inclusion initiatives

smart contracts due to intricate coding, decentralized collaboration, and potential oversights. To navigate these challenges, best practices have emerged, encompassing formal verification and standardized templates. While formal verification employs mathematical proofs to assure code correctness, it is a time-consuming process. Standardized templates, while reducing risks, might curtail customization possibilities.

Among the vulnerabilities that plague smart contracts, the following are salient (Fig. 5.2):

1. Re-entrancy Vulnerability: Unexpected Behaviour and Misuse of Funds

External contracts can exploit this vulnerability, potentially manipulating the contract's logic and misusing funds before the initial execution completes. Such attacks can lead to financial losses and data breaches [10, 11].

2. Integer Overflow/Underflow: Unforeseen and Malicious Outcomes

Arithmetic operations exceeding maximum or minimum values can result in unintended and potentially harmful consequences. Attackers might manipulate calculations to gain unauthorized access or cause contract failures.

3. Unauthorized Access: Manipulation of Critical Information

Inadequate access restrictions to sensitive functions or data can allow unauthorized parties to manipulate crucial information, undermining the confidentiality and integrity of contract operations.

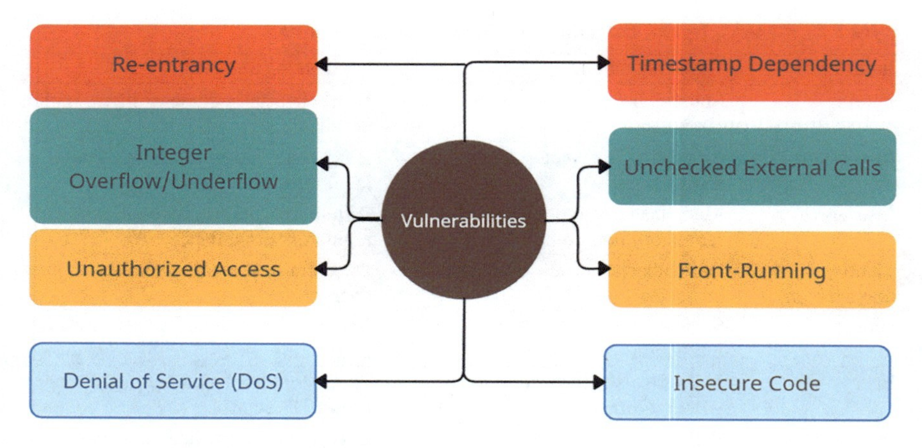

Fig. 5.2 Vulnerabilities in smart contracts

4. Denial of Service (DoS): Contract Unavailability and Slowdown

Exploiting vulnerabilities within smart contracts can lead to excessive resource consumption, causing unavailability or significant slowdown. This disrupts the contract's intended functionality.

5. Timestamp Dependency: Manipulation of Contract Logic

Contracts heavily reliant on block timestamps can be susceptible to manipulation by miners, potentially distorting the execution logic and leading to unexpected outcomes [11].

6. Unchecked External Calls: Exposure to External Malicious Contracts

Failure to validate and sanitize data from external contracts can expose a smart contract to vulnerabilities, especially when interacting with malicious external contracts.

7. Front-Running: Exploitation of Transaction Timing

Unscrupulous individuals exploit timing gaps between transaction submissions and blockchain inclusion to manipulate contract status, compromising fairness and security.

8. Insecure Code: Vulnerabilities Leading to Compromised Contracts

Code vulnerabilities like buffer overflows, SQL injections, or improper input validation can compromise the security of smart contracts, enabling attackers to compromise contract integrity [6].

Addressing these vulnerabilities demands meticulous code audits, rigorous testing, and robust security practices. Professionals experienced in smart contract development are crucial in this process, considering the nuanced intricacies of the blockchain environment.

However, the lack of standardized security protocols poses a significant challenge. Despite emerging best practices, the absence of consensus results in confusion, inconsistency, and security gaps during the evolution phase.

In summation, secure smart contract development encounters challenges stemming from intricate coding, decentralized collaboration, and potential oversights. Best practices such as formal verification and standardized templates offer mitigation strategies [12]. Noteworthy vulnerabilities encompass re-entrancy, integer overflow/underflow, unauthorized access, DoS attacks, timestamp dependency, unchecked external calls, front-running, lack of access controls, poor randomness, and insecure code. To ensure the credibility and reliability of blockchain technology, adept professionals must conduct thorough audits and testing. Establishing standardized security protocols is pivotal in achieving consistent security across smart contracts, enhancing their potential to transform industries while maintaining robust security measures.

6 Conclusion

This comprehensive study delves into the intricate landscape of smart contracts, shedding light on their profound significance, vulnerabilities, and applications across a wide array of industries. This journey reveals several overarching themes that underscore the multifaceted nature of smart contracts within the realm of blockchain technology.

An Integrated View of Smart Contracts.

Smart contracts offer the allure of efficiency, transparency, and automation, but they also present a multifaceted tapestry where technology and economics intertwine. These digital agreements possess the potential to redefine traditional contract practices and foster innovation across industries. Yet, to fully comprehend their transformative capacity, it's crucial to embrace a comprehensive perspective that not only acknowledges their potential but also acknowledges the challenges they pose.

Overcoming Challenges for a Transformed Tomorrow.

The path toward harnessing the complete potential of smart contracts is strewn with challenges. Our exploration of vulnerabilities, spanning from re-entrancy issues to insecure code, underscores the necessity for meticulous code audits, rigorous testing, and robust security measures. Furthermore, our journey across various sectors exposes the unique challenges each industry encounters when integrating and adopting smart contracts, including regulatory barriers and intricacies of integration.

Striking the Right Balance Between Innovation and Realism.

As we navigate the future landscape shaped by smart contracts, achieving a delicate equilibrium between innovation and pragmatism becomes paramount. While the allure of automation and efficiency is compelling, it must be accompanied by a careful approach that addresses security vulnerabilities and respects the distinct regulatory frameworks of each sector. The desire for customization must be harmonized with the imperative for standardized security protocols.

In the grand tapestry of technological progress, smart contracts emerge as dynamic threads interweaving innovation and challenges. Their potential to revolutionize industries is indisputable, yet so are the vulnerabilities they introduce. By embracing a holistic viewpoint, confronting challenges collectively, and striking the optimal balance between innovation and pragmatism, we stand at the precipice of a transformed future. Smart contracts are more than just a technological concept; they serve as a testament to our capacity to harness technology's potential while upholding its integrity. As industries continue to evolve and smart contracts refine their role, this journey signifies the dynamic interplay between human ingenuity and the ever-evolving digital landscape.

References

1. Hewa, T. M., Hu, Y., Liyanage, M., Kanhare, S., & Ylianttila, M. (2021). Survey on blockchain-based smart contracts: technical aspects and future research. *IEEE Access, 9*, 87643–87662.
2. Badi, S., Ochieng, E. G., Nasaj, M., & Papadaki, M. (2020). Technological, organisational and environmental determinants of smart contracts adoption: UK construction sector viewpoint. *Construction Management and Economics, 39*, 36–54.
3. Wang, S., Ouyang, L., Yuan, Y., Member, S., Ni, X., Han, X., & Wang, F.-Y. (2019). Blockchain-enabled smart contracts: Architecture, applications, and future trends. *IEEE Transactions on Systems, Man, and Cybernetics and Systems, 49*(11), 2266–2277.
4. Idrees, S., & Nowostawski, M. (2022). *Transformations through blockchain technology*. Springer.
5. Kemmoe, V. Y., Stone, W., Kim, J., Kim, D., & Son, J. (2020). Recent advances in smart contracts: a technical overview and state of the art. *IEEE Access, 8*, 117782–117801.
6. He, D., Deng, Z., Zhang, Y., Chan, S., Cheng, Y., & Guizani, N. (2020). Smart contract vulnerability analysis and security audit. *IEEE Network, 34*, 276–282.
7. Destefanis, G., Marchesi, M., Ortu, M., Tonelli, R., Bracciali, A., & Hierons, R. M. (2018). *Smart contracts vulnerabilities: A call for blockchain software engineering?* (pp. 19–25). 2018 International Workshop on Blockchain Oriented Software Engineering (IWBOSE).
8. Destefanis, G., Marchesi, M., Ortu, M., Tonelli, R., Bracciali, A., & Hierons, R. (2018). Smart contracts vulnerabilities: A call for blockchain software engineering. In *Proc. IEEE 1st Int. Work. Blockchain Oriented Softw. Eng* (pp. 19–25). https://doi.org/10.1109/IWBOSE.2018.8327567
9. Mehar, M. I., Shier, C. L., Giambattista, A., Gong, E., Fletcher, G., Sanayhie, R., Kim, H. M., & Laskowski, M. (2017). Understanding a revolutionary and flawed grand experiment in blockchain: The DAO attack. *Banking & Insurance eJournal.*
10. Dingman, W., Cohen, A., Ferrara, N., Lynch, A., Jasinski, P., Black, P. E., & Deng, L. (2019). Defects and vulnerabilities in smart contracts, a classification using the Nist bugs framework. *The International Journal of Networked and Distributed Computing, 7*, 121–132. https://doi.org/10.2991/ijndc.k.190710.003
11. Zheng, Z., Xie, S., Dai, H., Chen, W., Chen, X., Weng, J., & Imran, M. A. (2019). An overview on smart contracts: challenges, advances and platforms. *Future Generation Computer Systems, 105*, 475–491.
12. Singh, A., Parizi, R. M., Zhang, Q., Choo, K. K. R., & Tanha, A. D. (2020). Blockchain smart contracts formalization: Approaches and challenges to address vulnerabilities. *Computers & Security, 88*, Art. no. 101654. https://doi.org/10.1016/j.cose.2019.101654

13. Parjuangan, S., & Suhardi. (2020). *Systematic literature review of blockchain based smart contracts platforms* (pp. 381–386). 2020 International Conference on Information Technology Systems and Innovation (ICITSI).
14. Wang, S., Yuan, Y., Wang, X., Li, J., Qin, R., & Wang, F. (2018). *An overview of smart contract: architecture, applications, and future trends* (pp. 108–113). 2018 IEEE Intelligent Vehicles Symposium (IV).

Chapter 6
Modernizing Healthcare Data Management: A Fusion of Mobile Agents and Blockchain Technology

Ashish Kumar Mourya, Gayatri Kapil, and Sheikh Mohammad Idrees

1 Introduction

Blockchain technology offers a significant advantage in the form of a timestamp protocol within distributed networks. The peer-to-peer architecture holds its own importance, wherein each node functions as both a server and a client. Information within the network can be shared and broadcasted among connected nodes, with authentication playing a pivotal role. A ledger retains comprehensive logs detailing past and future events, while its interconnectivity across domains provides substantial advantages over alternative systems. A prime example, Bitcoin's blockchain technology enables over 10,000 computer nodes to process transaction requests, ensuring visibility to all connected nodes. Notably, the peer-to-peer structure eliminates the vulnerability of a single point of network failure [1–3].

In the realm of healthcare, an updated and decentralized database assumes a crucial role in facilitating interactions among diverse users seeking access to stored information across shared networks [4–6]. Ethereum's blockchain technology, in conjunction with a US start-up program, has given rise to the Gem Health Network. This innovative approach has proven pivotal within the healthcare sector, enabling seamless access to shared information concerning various diseases and patients by healthcare specialists. This ecosystem platform presents an opportunity for doctors, businesses, and experts to collaborate effectively and promptly. Further groundbreaking initiatives in healthcare, such as the collaboration between Guardtime and

A. K. Mourya (✉) · G. Kapil
Department of Computer Science, GNIOT Institute of Professional Studies, Greater Noida, India

S. M. Idrees
Department of Computer Science, Norwegian University of Science and Technology, Gjovik, Norway

S. M. Idrees, M. Nowostawski (eds.), *Blockchain Transformations*, Signals and Communication Technology, https://doi.org/10.1007/978-3-031-49593-9_6

Estonia, emphasize the utilization of blockchain technology for patient and physician operations.

In traditional databases, a central node connects to all other nodes, and if this central node were to fail, the entire network would go down. However, in a blockchain, no central node exists; instead, nodes are interconnected in a chain-like structure. This design means that if one node fails, the overall network remains unaffected. While traditional databases offer functions like read, write, and delete – which are crucial in cases such as insurance claims – blockchain operates differently. Data in a blockchain cannot be altered; it can only be created or read. Consequently, the risk of fraud through data modification is eliminated. Moreover, traditional databases often incur high costs, discouraging the storage of redundant data. In contrast, each block in a blockchain contains the entire dataset, ensuring zero data loss. This characteristic renders it suitable for applications requiring comprehensive record-keeping. The electronic storage nature of blockchain also ensures data safety, preventing physical damage or misplacement. Every transaction within a blockchain is time stamped, recording the creation and modification date and time of each block. This feature simplifies the identification of re-targeted blocks.

In a traditional database system, administrators can alter data ownership. Conversely, in blockchain, only the block's owner can change ownership, and such changes are traceable. This attribute enhances the storage of critical information. Furthermore, blockchain secures data through cryptography. Information is stored in encrypted form, accessible only to individuals possessing the private key. This approach guarantees network information security and ensures that only authorized individuals can access it [7, 8].

Mobile agents are dynamic entities that operate within distributed networks. They possess the ability to move freely within the system, pausing and resuming operations at will. Even if network connectivity is lost, mobile agents retain the capacity to roam and return to their source once connectivity is reestablished. Therefore, mobile agents can effectively operate within a blockchain network to access data without being affected by node connectivity disruptions.

Cloud computing constitutes a pooled resource and employs a multi-tenancy architecture for storage, connectivity, and capacity. However, this architecture's privacy risk encompasses potential failures in mechanisms for segregating storage, memory, routing, and reputation among various users sharing the infrastructure. The centralization of backups, storage, and pooled data warehouse space exposes centralized computing users to a higher risk of unauthorized access and leakage of private and sensitive information [10].

2 Related Works

Minimizing costs and enhancing facilities have become a crucial organizational requirement. A cloud-based medical data processing platform was proposed by [11], employing Big Data processing tools. This initiative involved the storage of data within a cloud environment, aggregating information from distributed nodes. The inventor utilized web-based search techniques to facilitate data retrieval. Additionally, patient data analysis was conducted to discern patterns and diagnose diseases with remote expert assistance. In the realm of healthcare industries, the generation of multi-source unstructured data has posed challenges. Addressing the heterogeneity prevalent in unstructured data, [12] delved into the domain of telemedicine. Their solution introduced a novel architecture aimed at collecting patient diagnostic information within a cloud-based application, subsequently storing this data for future predictive analysis.

Further contributions include the proposition of an E-health application based on a cloud platform by [13], termed Remote Telemetry Units. This application leverages IoT-based sensor techniques to capture data and employs Big Data methodologies for processing medical information. A noteworthy approach, put forth by researchers in [15], introduces a context-aware model capable of delivering multiparty services within diverse and mobile environments. This architectural model boasts the ability to dynamically switch networks according to current conditions, ensuring the seamless delivery of multimedia streaming content across varied networks.

These researchers offered a proficient context-aware system to minimize energy consumption in electronic communication devices [16]. They have designed a simulated battery consumption model funded by Nokia Energy Profiler tool to investigate the battery consumption based on multiple sensing devices configuration and background. With the simulation model, context-aware approaches examine to resolve and optimize energy consumption at various levels.

The patient's data collection information application model has been proposed by [17] in healthcare organization based on cloud computing technology, which exchanges the health-related data in the cloud storage for processing and distribution. This proposed model is still sprouting, but this model supports and provides facilities for one client or single institutional access and normally has a massive cost to integrate computing, information, or communication exchanges from outside of the cloud, particularly the mobile segment of the overall system.

In healthcare sector, agent-based techniques are deploying exponentially, although, there are several agent-based solutions for interoperability among health information system [18]. Isern et al. [19] propose multi-agent system architecture and knowledge representation technique to allow the enactment of clinical guidelines. Lanzola [20] introduces a methodology for the expansion of interoperable agents to be implemented in healthcare model. Tyson et al. [21] propose an agent-based application for employment in clinical trials, and these researchers demonstrate that the agent-based model has numerous potentialities over the centralized architecture. Kim et al. [22] proposes an approach for designing and deploying union mobile agents (UMAs) to extend ubiquitous healthcare. Kaluža [23] divulges a multi-agent system that serves aged people, checking falls and topping out deficiencies through an intelligent environment augmented by sensing devices.

3 Mobile Agents for Blockchain Technology

Mobile agents are run-time entities based on software agents that work on distributed networks. These agents play an important role on unreliable networks. The deployment and usage of agents provided many advantages like diminution of network traffic and latency, fault tolerance, inducing robustness, filtering, and remote searching. It has the ability of being autonomous and moves freely crosswise in an uninformed system, halts, and resumes at any time. The mobile agents can work in an asynchronous mode. These characteristics of mobile agents make them efficient to work on unreliable and low bandwidth networks. Even if networks lost their connectivity, mobile agents have sufficient privileges to work on roam and come back to their source once node got connection. Mobile agents have the ability to migrate from one location to another. They utilize less bandwidth on Internet. Mobile agents have been used in e-commerce business to save time and eliminating fraud transaction on a network [9]. Many mercantile and electronic transactions require access to assets in real time. The capability of a mobile agent to incarnate their creator's intentions and to act and bargain on behalf of them makes it well suited for online trading. We have used mobile agents to store information send them to the different clients who needs it.

Mobile agents have the capability to move over the distributed networks. We have proposed a framework where a mobile agent fetches the prediction of disease-related information over the networks and spread them over the connected nodes. Researchers currently use blockchain technology and machine learning techniques. This kind of architecture has a significant role where hybrid technology can be used for providing a new solution to the medical industries. The purpose of machine learning with blockchain technology and mobile agents is a new concept. Every single processed information through machine learning techniques will be spread over the broadcast network, and clients requiring access to specific information can retrieve it through authentication. The mobile agents also have durability. If a client has been disconnected from network, the mobile agents will be delivered the information after the client got connectivity again over the network. One major advantage of using mobile agents with blockchain technology is that it works as an agent [14].

In the healthcare industry, the integration of mobile agents and blockchain technology holds promising potential for enhancing pharmaceutical processes. Within the pharmaceutical sector, the utilization of agents becomes invaluable for tasks ranging from promotional endeavors to efficient drug delivery. By leveraging mobile agents, we can streamline drug delivery logistics and establish robust traceability mechanisms. Moreover, this approach offers a viable solution to curbing the challenge of counterfeit drugs, ensuring the integrity of pharmaceutical supply chains. Furthermore, the convergence of mobile agents and blockchain technology opens the door to novel collaborations among healthcare professionals, researchers, and patients. This collaborative environment facilitates tailored and individualized patient care. The seamless maintenance and accessibility of electronic health records within this network of healthcare providers can lead to improved prescription accuracy and holistic solutions for patients.

In the ever-evolving landscape of healthcare, the application of machine learning techniques for disease prediction stands as a pivotal focus. The synergy between mobile agents and blockchain technology enhances this facet by contributing to the accuracy of predictive models and facilitating the integration of diverse data sources. The healthcare and medical diagnosis domains have borne witness to the substantial impacts of mobile agents and blockchain technology, underscoring their significance in revolutionizing patient care and diagnostic practices.

4 Proposed Model Based on Mobile Agents

Blockchain technology is a peer-to-peer networking platform that works with the Internet protocols. Blockchain has ability to store and records every transaction, which has completed the entire necessary requirement in the distributed environment. The information is stored in a shared platform, but its content may be accessible through permission or privilege right to gathered data. The blockchain technology has its own protocols that contain log information about client, timestamp, and crypto-graphical information. The digital signature makes the authentication process more secured. If the networking protocol has been broadcasted over the network, it can be visible to all the connected nodes but accessible by those for which it has been generated. These information are generated in form of block, and it will be added with other blocks that are already available in the form of logs.

This endeavor introduces a healthcare database processing system rooted in blockchain technology, fortified with the capabilities of mobile agents. These agents exhibit a range of qualities, including cross-platform compatibility and adaptability to dynamic heterogeneous environments. This amalgamation aims to establish a secure and dependable ecosystem for processing healthcare information, catering to Medicare services. The resulting architecture significantly enhances the convenience of health record searches and data queries, offering a seamless and efficient experience.

This framework comprises three distinct components. The first segment encompasses the Client Authentication process, ensuring secure user verification. The second segment revolves around the migration of mobile agents, showcasing their dynamic movement within the system. The final component elucidates how the cloud facilitates the storage and processing of healthcare information, underscoring its pivotal role in the framework.

4.1 Client Authentication Process

The Client Authentication process assumes a pivotal role within the proposed architecture, serving as a cornerstone. Diverse user categories, including Doctors, Patients, and medical service users, benefit from the cloud environment, enabling seamless information storage and access from any location. This accessibility is facilitated through individualized login credentials, ensuring secure entry points. The System Administrator wields comprehensive privileges, retaining the authority to modify or revoke rights if users fail to meet prerequisites or engage in unauthorized activities, thereby safeguarding the system's integrity.

The verification procedure entails the allocation of a unique identifier to clients, fortified by encryption and decryption technologies, bolstering authentication. Central to this mechanism is the registration of all mobile agent systems within the primary server, establishing a controlled network. Exclusive task assignment becomes possible post thorough client verification and validation, thus maintaining stringent access controls (Figs. 6.1 and 6.2).

4.2 Migration of the Mobile Agents

In this envisioned architecture, clients leverage web-based mobile agents to process healthcare information sourced from external origins and remote locations, as depicted in Fig. 6.3. Before a mobile agent embarks on its designated task, it must

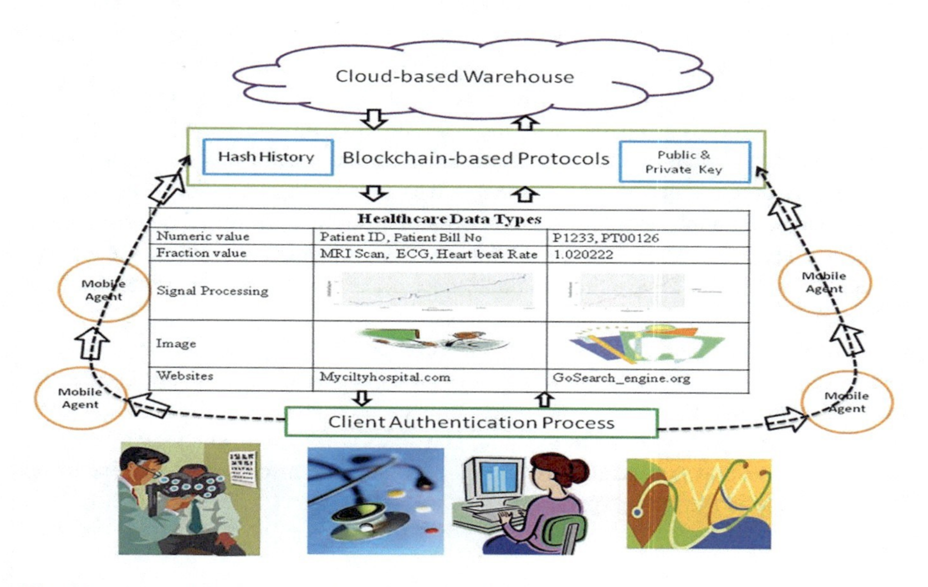

Fig. 6.1 Proposed blockchain technology with mobile agents

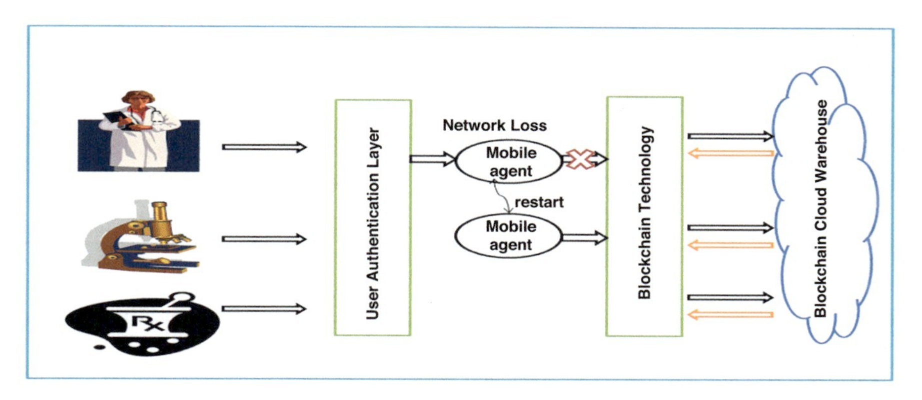

Fig. 6.2 Crawling process of mobile agents even after network losses

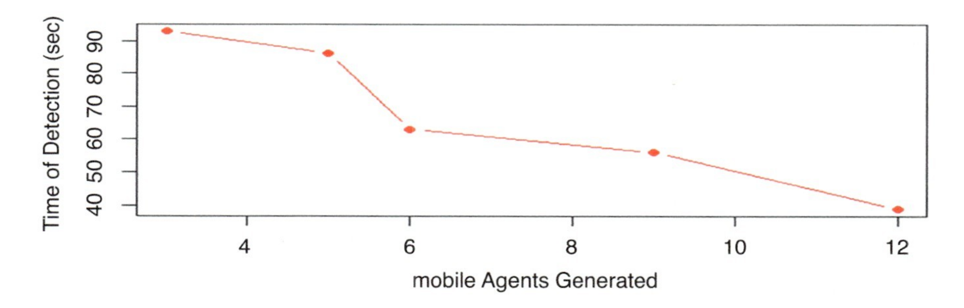

Fig. 6.3 Time detection with mobile agents

undergo a stringent authentication process. Furthermore, the mobile agent is required to carry its authentication throughout its lifecycle as it traverses from one system to another, ensuring seamless acquisition of client information. Initially, mobile agents engage in data collection, extracting information from their respective clients, and subsequently transmitting it to a cloud-based repository utilizing blockchain technology. Within this architecture, each node maintains an individual hash history, contributing to enhanced security measures. Transactions are securely executed through data encapsulation techniques, with a significant role played by the public and private key mechanism, reinforcing data integrity.

Consider a scenario in which a patient possessing a pre-existing medical history seeks admission to a specific Outpatient Department (OPD) within a particular clinic. The specialist doctor within that department exercises their access privileges, entrusting a mobile agent with the task of retrieving the patient's medical history. Upon receiving the assigned task, the mobile agent initiates the collection of the patient's historical data from other hospital systems or collaborates with other mobile agents based on the prevailing strategy. Notably, the mobile agent retains its operability even in instances of low or disrupted communication network speeds, thereby rendering it suitable for environments susceptible to information loss, as illustrated in Fig. 6.2. This resilience underscores the adaptability of the mobile agent in challenging communication conditions.

4.3 Warehouse Based on the Cloud Technology

The cloud computing environment is comprised of an array of interconnected servers hosted online via the Internet. These servers offer organizations a range of services, including data storage, management, and processing. However, the reliance on cloud service providers introduces significant security concerns. Cloud technology provides a shared, customizable, and easily accessible resource pool, requiring minimal effort for interaction and management.

Within the healthcare sector, the utilization of a cloud-based warehouse holds immense significance. Both public and private cloud configurations can be harnessed for this purpose. Clients gain the ability to access data remotely, tailored to their specific requirements. Each node within the proposed model adheres to decentralized protocols, ensuring a distributed and resilient network.

Nodes associated with patients store a comprehensive set of data, encompassing prescriptions, medical history, family background, test results, and medications. This data can be shared with insurance providers for insurance claims and with doctors for accurate treatment. Simultaneously, insurance providers maintain datasets that encompass policy details, government regulations, insured individuals' particulars, and financial records.

Furthermore, domain expert nodes house intricate patient medication details. The proposed model establishes robust connectivity among critical components within the healthcare sector, yielding substantial benefits to the entire industry. As a cohesive and interconnected framework, this model holds the potential to revolutionize the healthcare landscape.

5 Results and Experiments

The proposed research has used a private Ethereum network with a custom genesis block to allocate ethers crypto currency to all the mobile agents. On the other hand, for mobile agents, [27] has been used to emulate the network of nodes. These mobile agents have traveled into the network with blockchain transaction and stored information about the generated route.

The experiment has been performed using a fully connected network created using different nodes [27]. From Table 6.1, we have observed that on an average, as the agent increases, the time consumed for detection of fault decreases. This is because with more mobile agents the chances of reporting must be high. Due to randomly generating characteristics of mobile agent, detection time will be not constant. The proposed study has noticed that there was a variance between number of

Table 6.1 Detection time and transaction time

Mobile agents generated	Time of detection (sec)	Block size	Transaction time
3	93	300	20
5	86	600	36
6	63	900	43
9	56	1200	59
12	39	1500	66

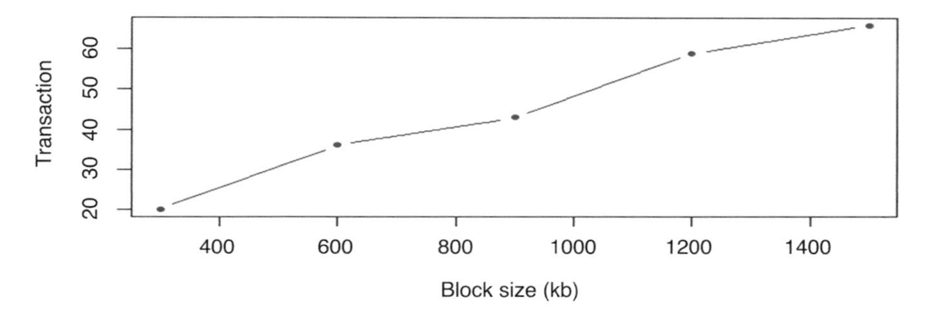

Fig. 6.4 Transaction time of blockchain

mobile agents and time of detection (Fig. 6.3). The detection time was low with the maximum numbers of agents. On the other hand, there was sudden increase with the maximum number of blocks (Fig. 6.4).

5.1 Algorithm for Generating Blockchain

This study has used R Studio for generating blockchain for simulating our proposed work.

```
Library(Rbitcoin)


now=Sys.time()


blocknumber_1 <- list(number = 1,
 timestamp = now,
 data = "Ashish",
 parent = NULL)

blocknumber_2 <- list(number = 2,
 timestamp = now,
 data = "Shafqat",
parent_address = 1)

blocknumber_3 <- list(number = 3,
 timestamp = now,
 data = "Majid",
parent_address = 2)

Implement_blockchain = list(blocknumber1, blocknumber2,
blocknumber3)


Call_blockchain[[2]]
```

5.2 Algorithm for Generating Hash for Blockchain

```
hash_block<- function(block)
{
block$new_hash<- digest(c(block$index,
block$timestamp,
block$data,
block$previous_hash), "spookyhash")
return(block)
}
```

5.3 Algorithm for Generating and Migrating Mobile Agent

```
public interface Mobile extends
 java.io.Serializable
```

```
 {
public void go(java.net.URL dest)
 throws
com.ibm.aglet.RequestRefusedException,
edu.ohio_state.cis.brew.MoveRefusedException,
java.io.IOException;...
 }
public class AgentImpl extends
 MobileObject implements Agent
 {
 int a;
 public AgentImpl()
 {/* init code */}
 public void foo(int x)
 throws AgletsException
 { BC; }
 }
protected class Foo extends Frame
{
int x, y, progCounter = 0; Object trgt;
void setPCForMove()
{ ... }
void run()
{
try
{ ...AgentImpl.this.request_read();
if ((progCounter == 0))
{
progCounter+=1; BC1
 }
AgentImpl.this.read_accomplished();
AgentImpl.this.request_read();
 if ((progCounter == 1))
{
progCounter+=1; BC2
}
AgentImpl.this.read_accomplished();
 } catch(AgletsException e)
{ ... }
} ... }
```

5.4 Comparison of Proposed Model with Existing Models

The proposed approach exhibits notable advantages in comparison to previous

Table 6.2 Comparison of proposed model

Research work	Type of data used	Challenge addressed	Fault tolerance	Scalability	Authentication
Zhang et al. [24]	Electronic health records	Interoperability, access control, data integrity	No	No	Yes
Shan et al. [25]	Electronic health records	Access control, interoperability	No	Yes	No
Tanwar et al. [26]	Electronic health records	Access control, interoperability	No	No	Yes
Proposed model	Electronic health records	Interoperability, access control, data integrity, fault tolerance, reconnection	Yes	Yes	Yes

models, as depicted in Table 6.2. This study undertook a comprehensive comparison, considering aspects such as interoperability, data integrity, and fault tolerance. The standout attribute of our proposed model is its exceptional fault tolerance capability. Mobile agents possess the inherent ability to regenerate from where they were paused or even reconstitute themselves in the event of destruction caused by disconnectivity. This inherent resilience enhances the robustness of the proposed model.

6 Conclusion

In conclusion, blockchain technology presents a promising solution to address challenges within the medical and healthcare sectors, exemplified by its utilization in e-health records and nationwide data interoperability. Nonetheless, the implementation of this technology on a larger scale within healthcare necessitates further research and experimentation to establish a secure foundation. While the theoretical potential of blockchain has been acknowledged in recent years, its practical application is still evolving. The significance of data cannot be overstated; it pervades every aspect of our lives, enabling us to predict individuals' medical histories and conditions. This study introduces a novel framework employing blockchain technology and mobile agents to enhance prediction markets and the healthcare industry. The amalgamation of blockchain and mobile agent architecture offers heightened security and durability. By aggregating predictive outcomes from diverse nodes through mobile agents, a resilient structure is achieved. Mobile agents, with their ability to traverse distributed networks, serve as a conduit for storing machine learning results. Their autonomy to move within the system, pause, and resume operations, even during network disruptions, underscores their resilience. Thus, integrating mobile agents into the blockchain network facilitates uninterrupted data access, resilient to node connectivity disruptions.

Future endeavors will focus on minimizing deployment costs and streamlining network synchronization challenges. The symbiotic characteristics of blockchain and mobile agents promise a robust and secure model that ensures data security mechanisms, authentication, and integrity checks for online data.

References

1. Longo, D. L., & Drazen, J. M. (2016). Data sharing. *The New England Journal of Medicine, 374*(3), 276–277. https://doi.org/10.1056/NEJMe1516564
2. Almashaqbeh, G., Hayajneh, T., Vasilakos, A. V., & Mohd, B. J. (2014). Qos-aware health monitoring system using cloud-based wbans. *Journal of Medical Systems, 38*(10), 121.
3. Hayajneh, T., Mohd, B. J., Imran, M., Almashaqbeh, G., & Vasilakos, A. V. (2016). Secure authentication for remote patient monitoring with wireless medical sensor networks. *Sensors, 16*(4), 424.
4. Fan, K., Wang, S., Ren, Y., Li, H., & Yang, Y. (2018). MedBlock: Efficient and secure medical data sharing via blockchain. *Journal of Medical Systems, 42*(8), 136. https://doi.org/10.1007/s10916-018-0993-7
5. Mourya, A. K., & Idrees, S. M. (2020). Cloud computing-based approach for accessing electronic health record for healthcare sector. In *Microservices in Big Data Analytics: Second International, ICETCE 2019, Rajasthan, India, February 1st-2nd 2019, Revised Selected Papers* (pp. 179–188). Springer Singapore.
6. Idrees, S. M., Nowostawski, M., Jameel, R., & Mourya, A. K. (2021). Privacy-preserving infrastructure for health information systems. In *Data Protection and Privacy in Healthcare* (pp. 109–129). CRC Press.
7. Kuo, T.-T., Kim, H.-E., & Ohno-Machado, L. (2017). Blockchain distributed ledger technologies for biomedical and health care applications. *Journal of the American Medical Informatics Association, 24*(6), 1211–1220.
8. Ciampi, M., et al. (2013). A federated interoperability architecture for health information systems. *Int'l Journal of Internet Protocol Technology, 7*(4), 189–202.
9. Eyal, I., & Sirer, E. G. (2018). Majority is not enough: bitcoin mining is vulnerable. *Commun. ACM, 61*(7), 95–102.
10. Angraal, S., Krumholz, H. M., & Schulz, W. L. (2017). Blockchain technology: Applications in health care. *Circulation: Cardiovascular Quality and Outcomes, 10*(9), e003800.
11. Yang, C.-T., Liu, J.-C., Chen, S.-T., & Lu, H.-W. (2017). Implementation of a big data accessing and processing platform for medical records in cloud. *Journal of Medical Systems, 41*(10).
12. Idrees, S., & Nowostawski, M. (2022). *Transformations through blockchain technology*. Springer.
13. Suciu, G., Suciu, V., Martian, A., Craciunescu, R., Vulpe, A., Marcu, I., Halunga, S., & Fratu, O. (2015). Big data, internet of things and cloud convergence – An architecture for secure e-health applications. *Journal of medical systems, 39*(11), 141.
14. Mourya, Kumar, A., & Singhal, N. (2014). *Managing congestion control in mobile AD-HOC network using mobile agents*. ArXiv: 1401.4844 [Cs], January 20, 2014. http://arxiv.org/abs/1401.4844
15. Antoniou, J., Christophorou, C., Janneteau, C., Kellil, M., Sargento, S., Neto, A., Pinto, F. C., Carapeto, N. F., & Simoes, J. (2009). *Architecture for context-aware multiparty delivery in mobile heterogeneous networks* (pp. 1–6). ICUMT '09. International Conference on Ultra Modern Telecommunications & Workshops.
16. Bernal, J. F. M., Ardito, L., Morisio, M., & Falcarin, P. (2010). *Towards an efficient context-aware system: Problems and suggestions to reduce energy consumption in mobile devices*

(pp. 510–514). Ninth International Conference on Mobile Business and Global Mobility Roundtable (ICMB-GMR).
17. Rolim, C. O., Koch, F. L., Westphall, C. B., Werner, J., Fracalossi, A., & Salvador, G. S. (2010). *A cloud computing solution for patient's data collection in health care institutions* (pp. 95–99). ETELEMED '10. Second International Conference on eHealth, Telemedicine, and Social Medicine.
18. Isern, D., Sánchez, D., & Moreno, A. (2010). Agents applied in healthcare: A review. *International Journal of Medical Informatics, 79*, 145–166.
19. Isern, D., Sánchez, D., & Moreno, A. (2012). Ontology-driven execution of clinical guidelines. *Computer Methods and Programs in Biomedicine, 107*, 122–139.
20. Lanzola, G., Gatti, L., Falasconi, S., & Stefanelli, M. (1999). A framework for building cooperative software agents in medical applications. *Artificial Intelligence in Medicine, 16*, 223–249.
21. Tyson, G., Taweel, A., Miles, S., Luck, M., van Staa, T., & Delaney, B. (2012). An agent-based approach to real-time patient identification for clinical trials. In *Electronic healthcare* (Vol. 91, pp. 138–145). Springer.
22. Kim, H. K. (2014). Convergence agent model for developing u-healthcare systems. *Future Generation Computer Systems, 35*, 39–48.
23. Kaluža, B., Cvetkovic, B., Dovgan, E., Gjoreski, H., Gams, M., Luštrek, M., & Mirchevska, V. (2013). A multi-agent care system to support independent living. *International Journal on Artificial Intelligence Tools, 23*. https://doi.org/10.1142/S0218213014400016
24. Zhang, P., White, J., Schmidt, D. C., Lenz, G., & Rosenbloom, S. T. (2018). FHIRChain: Applying blockchain to securely and scalably share clinical data. *Computational and Structural Biotechnology Journal, 16*, 267–278.
25. Shan, J., Jiannong, C., Hanqing, W., Yanni, Y., Mingyu, M., & Jianfei, H. (2018). *Bloc HIE: a BLOCkchain-based platform for healthcare information Exchange, 2018 IEEE International Conference on Smart Computing (SMARTCOMP), 18–20 June 2018*. IEEE Computer Society.
26. Tanwar, S., Parekh, K., & Evans, R. (2020). Blockchain-based electronic healthcare record system for healthcare 4.0 applications. *Journal of Information Security and Applications, 50*, 102407.
27. Semwal, T., Nikhil, S., Jha, S. S., & Nair, S. B. (2016). *Tartarus: A multi-agent platform for bridging the gap between cyber and physical systems (demonstration)*. AAMAS.

Chapter 7
Machine Learning Approaches in Blockchain Technology-Based IoT Security: An Investigation on Current Developments and Open Challenges

P. Hemashree, V. Kavitha, S. B. Mahalakshmi, K. Praveena, and R. Tarunika

1 Introduction

In the contemporary interconnected landscape, merging blockchain technology with the Internet of Things (IoT) has surfaced as a potential remedy to tackle the significant security challenges linked with IoT systems. As the adoption of IoT platforms continues to grow across various domains, including healthcare, smart cities, and industrial automation, the necessity for strong security measures is of utmost importance. The decentralized and immutable nature of blockchain technology offers a potential solution by providing a tamper-resistant and transparent framework for data integrity and transaction verification.

However, despite the inherent security advantages of blockchain, IoT systems still face numerous challenges, ranging from device vulnerabilities and data integrity to privacy concerns and scalability issues. To further enhance the security posture of blockchain-based IoT systems, researchers have turned to machine learning algorithms as a complementary approach. By leveraging the power of machine learning, these systems can learn from vast amounts of data, detect anomalies, and identify potential threats in real time, ultimately strengthening the overall security framework.

This chapter aims to investigate the current developments and open challenges in the intersection of machine learning, blockchain technology, and IoT security. We will begin by providing an outline of blockchain technology, IoT, and the role of machine learning algorithms in enhancing security. Understanding the

P. Hemashree · S. B. Mahalakshmi · K. Praveena · R. Tarunika
Department of AI&ML, Coimbatore Institute of Technology, Coimbatore, Tamilnadu, India

V. Kavitha (✉)
Department of CGS, Sri Ramakrishna College of Arts and Science,
Coimbatore, Tamilnadu, India
e-mail: kavitha@srcas.ac.in

S. M. Idrees, M. Nowostawski (eds.), *Blockchain Transformations*, Signals and Communication Technology, https://doi.org/10.1007/978-3-031-49593-9_7

fundamentals of these technologies will set the stage for exploring their integration and the potential benefits they bring to IoT security.

Next, we will delve into the existing literature to identify relevant articles that have employed machine learning-based blockchain technology in IoT security. By reviewing these studies, we can gain insights into the practical implementation, effectiveness, and limitations of the proposed approaches. This analysis will provide a comprehensive understanding of the cutting-edge methods and their applicability in different IoT use cases.

Furthermore, this chapter will highlight the current developments in machine learning approaches for blockchain-based IoT security. We will examine real-world case studies and use cases that showcase the successful integration of machine learning algorithms with blockchain technology to mitigate security risks in IoT systems. By studying these implementations, we can better understand the potential benefits and challenges associated with deploying such solutions in practice.

Lastly, we will address the open challenges and identify potential improvements that can be focused on for further study. This includes scalability and performance issues, privacy concerns, interoperability, standardization, and ethical considerations. By acknowledging these challenges, we can provide insights for future research directions and pave the way for the adoption of machine learning approaches in securing IoT systems built on blockchain technology.

In summary, this chapter intends to provide a valuable contribution to the ongoing discussion regarding the intersection of machine learning, blockchain technology, and the security of IoT systems. By investigating the current developments, challenges, and potential improvements, we can provide researchers, practitioners, and policymakers with valuable insights and recommendations to enhance the security posture of IoT systems in the era of blockchain technology.

1.1 Background and Motivation

The rapid proliferation of IoT devices and the increasing reliance on blockchain technology for secure and transparent transactions have transformed various industries [51, 54]. However, the security vulnerabilities associated with IoT systems have posed significant challenges, making them attractive targets for cyberattacks [15]. The decentralized nature of blockchain technology presents an opportunity to improve the security of IoT systems by ensuring data integrity, immutability, and decentralized consensus [12, 40].

Machine learning, with its ability to learn patterns, detect anomalies, and make intelligent decisions from large datasets, has emerged as a powerful tool for enhancing security across various domains [35]. Within the framework of IoT systems based on blockchain technology, machine learning algorithms can play a vital role in detecting and mitigating potential threats, identifying patterns of behavior, and enhancing anomaly detection capabilities [13, 5].

The driving force behind this chapter is rooted in the necessity to delve into the intersection of machine learning, blockchain technology, and the security of IoT systems. By investigating the current developments and open challenges in this area, we aim to shed light on the potential benefits, limitations, and practical implications of integrating machine learning algorithms with blockchain technology for IoT security.

Understanding the current state of machine learning approaches in blockchain-based IoT security is crucial for researchers, practitioners, and policymakers to make informed decisions and develop effective security strategies [4]. This book chapter aims to connect the gap between theoretical concepts and practical implementations, providing insights into real-world use cases, case studies, and the associated challenges.

Moreover, this chapter seeks to identify potential improvements and research directions to address the limitations and open challenges in utilizing machine learning for enhancing the security of IoT systems built on blockchain technology. By addressing scalability, privacy, interoperability, and ethical considerations, we aim to foster the development of robust and secure solutions that can be readily implemented in IoT environments.

Overall, the background and motivation for this chapter stem from the need to explore and understand the current landscape of machine learning approaches in blockchain-based IoT security. By investigating the state-of-the-art techniques, identifying challenges, and providing future research directions, this chapter aims to contribute to the advancement of secure IoT systems and the utilization of machine learning algorithms in conjunction with blockchain technology.

1.2 Objectives and Scope

The objectives of this book chapter are as follows:

(i) To provide an outline of blockchain technology, IoT, and machine learning algorithms in the context of security, establishing the foundation for understanding the amalgamation of these technologies in IoT systems.
(ii) To investigate relevant articles and studies that have utilized machine learning-based blockchain technology for IoT security, examining their effectiveness, limitations, and practical implications.
(iii) To explore the current developments in machine learning approaches for blockchain-based IoT security, showcasing real-world case studies and use cases that highlight successful implementations.
(iv) To identify and discuss the open challenges associated with using machine learning in blockchain-based IoT security, including scalability, privacy concerns, interoperability, standardization, and ethical considerations.

(v) To provide recommendations for future research directions and potential improvements in the field, addressing the identified challenges and fostering the development of robust and secure solutions.

The scope of this chapter encompasses the intersection of machine learning, blockchain technology, and IoT security. It explores the current developments, challenges, and opportunities in utilizing machine learning approaches to improve the security of IoT systems built on blockchain technology.

This chapter focuses on examining the integration of machine learning algorithms with blockchain technology in the context of IoT security. It discusses the practical implementations, effectiveness, and limitations of these approaches through a review of relevant articles and studies. Real-world case studies and use cases are analyzed to understand the application of machine learning in securing IoT systems using blockchain technology.

While the chapter provides understanding of the most advanced techniques available, it also recognizes the open challenges and constraints that need to be addressed. It addresses scalability, privacy, interoperability, standardization, and ethical considerations as key challenges in deploying machine learning approaches in blockchain-based IoT security.

The chapter concludes with recommendations for future research, highlighting potential areas of improvement and suggesting avenues for further exploration to overcome the identified challenges.

2 Overview of Blockchain Technology and IoT Security

2.1 Blockchain Technology Fundamentals

Blockchain technology represents a decentralized and distributed ledger system designed to enable secure and transparent transactions between multiple participants, eliminating the necessity for intermediaries [40]. Figure 7.1 illustrates the formation of a blockchain using three blocks. It operates through a chain of blocks, with each block containing a set of verified and timestamped transactions [54]. The fundamental characteristics of blockchain technology include immutability, transparency, decentralization, and cryptographic security [11].

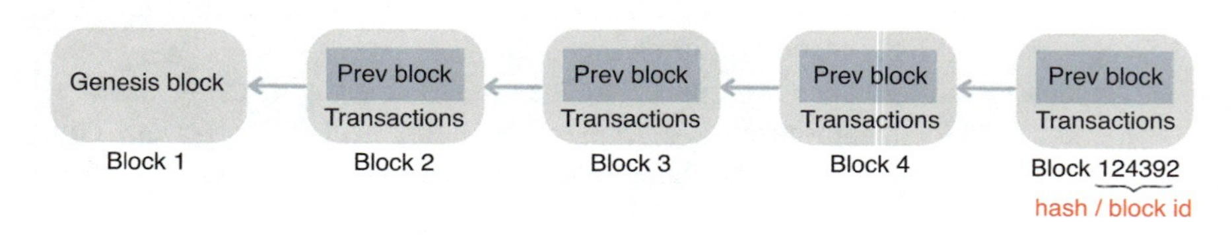

Fig. 7.1 Three blocks forming a chain

Consensus mechanisms play a crucial role in ensuring agreement on transaction validity and preventing malicious activities within a blockchain network. The Proof of Work (PoW) consensus mechanism, where participants contend to solve complex mathematical puzzles, is widely used in popular blockchain networks like Bitcoin [40]. Further, consensus mechanisms such as Proof of Stake (PoS) and Delegated Proof of Stake (DPoS) have gained traction due to their energy efficiency and scalability advantages [37, 63].

Smart contracts are programmable and automated contracts that can be deployed on blockchain networks. They operate based on predefined rules and conditions, enabling automation, reducing the need for intermediaries, and enhancing the efficiency and transparency of transactions [6].

Working of Blockchain

The sequence actions performed while working with blockchain [40] is as represented in Fig. 7.2.

The working of the blockchain is as follows:

(i) First under participants and nodes, there exists multiple participants that are often referred to as nodes, which join the network voluntarily. Every participating node possesses a duplicate of the complete blockchain ledger.
(ii) In the transaction creation, the participants initiate transactions by creating and digitally signing them. Transactions can represent various types of data such as financial transfer, contracts, assets ownership, or any other relevant information.
(iii) In the transaction propagation, once the transaction is created, it is transmitted across the network. Nodes receive the transaction and validate its authenticity and integrity.
(iv) In transaction validation, nodes independently verify the validity of each transaction using predefined rules and criteria. This validation process ensures that transactions comply with specific consensus rules, such as verifying digital signatures, checking available funds, or executing smart contract conditions.
(v) In pending transactions and mining, the validated transactions are added to a pool of pending transactions, waiting to be included in a new block. Miners, who are special nodes in the network, contest to solve a cryptographic puzzle or perform a consensus algorithm to secure the next block.
(vi) In block creation, the miners gather a set of pending transactions from the pool and create a new block. The block comprises a header and a list of transactions. The header encompasses metadata like a timestamp, a reference to

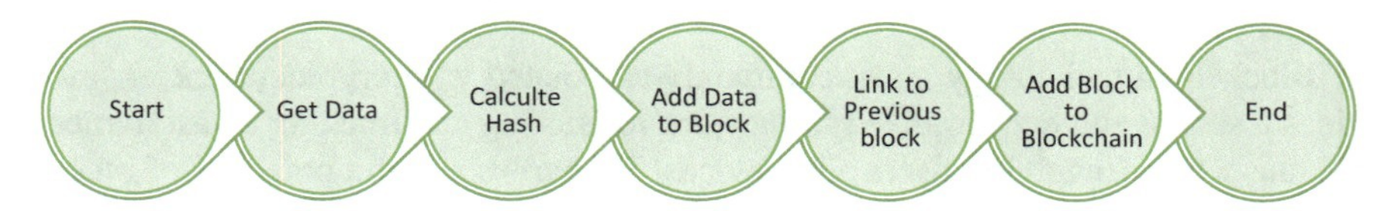

Fig. 7.2 Working of blockchain

the previous block's hash, and a nonce (a random number used in the proof-of-work process).

(vii) In consensus and block validation, the newly created block is propagated throughout the network. Other nodes verify the block's validity, including the correctness of the proof-of-work or consensus algorithm. If the block is valid, it is accepted and added to each node's copy of the blockchain.

(viii) In linking blocks, the accepted block becomes the latest addition to the blockchain. It includes a reference to the hash of the preceding block, effectively forming a chain of blocks. This linkage guarantees the immutability and truthfulness of the blockchain. Modifying the data in any block would require altering subsequent blocks and gaining control of the majority of the network's computational power.

(ix) In consensus and trust, as new blocks are added to the chain, consensus mechanisms ensure that all participants approve on the state of the blockchain. The decentralized nature of blockchain permits trust to be distributed among the participants rather than relying on a central authority.

(x) In data replication and synchronization, nodes in the network constantly share information to maintain a synchronized and consistent copy of the blockchain. Any changes to the blockchain, such as the addition of new blocks or updates to transaction history, are propagated across the network to keep all copies up to date.

2.2 *Internet of Things (IoT)*

Introduction to IoT

The Internet of Things (IoT) has gained significant prominence in various fields, transforming the way devices connect and communicate. However, as the adoption and advancement of IoT continue to accelerate, ensuring robust security has become a critical imperative. To address this challenge, the incorporation of blockchain technology with IoT has emerged as a possible solution, offering enhanced security and data integrity. This chapter aims to delve into the potential of blockchain in bolstering security within IoT systems.

IoT is a rapidly evolving technology that enables the seamless interconnection of diverse devices over the internet. Its development is an outcome of the convergence of technologies such as Radio Frequency Identification (RFID), wireless communication, and sensor networks. This architecture has been widely embraced by industries, unlocking new opportunities for revenue generation and operational efficiency [20].

Blockchain technology, most commonly associated with cryptocurrencies, provides a secure and decentralized framework for storing and tracking a vast number of transactions involving users and devices. By harnessing the potential of blockchain, IoT systems can benefit from enhanced security, privacy, and data integrity, mitigating potential vulnerabilities and threats [9].

The essence of IoT lies in the interconnectedness of various communication networks, allowing devices to interact and exchange data seamlessly through the internet. This connectivity brings forth numerous advantages, including improved automation, advanced data analytics, and operational efficiency, revolutionizing industries such as healthcare, manufacturing, and transportation [4].

Architectural View of IoT

The Internet of Things (IoT) encompasses a vast array of heterogeneous and large-scale terminal devices, facilitating their interconnectedness and communication. The architecture of IoT applications ideally consists of three layers: the things layer, the cloud layer, and the edge layer [4].

While the IoT architecture lacks a standardized structure, it is commonly categorized into either four or five layers. In the four-layered IoT architecture, the layers include the business layer, support layer, communication layer, and perception layer. The communication layer ensures the reliable transmission of information between different layers and encompasses sublayers such as application, session, transport, network, MAC, and physical layers. The support layer enhances the functionality of other layers by providing computing services and storage facilities, with fog/edge and cloud computing serving as key technologies within this layer. The business layer incorporates software applications that align with industry requirements and user specifications [52].

The things layer accommodates a diverse range of heterogeneous devices, including sensors and actuators. These end-terminal devices combine physical components that interact with the physical world and cyber components that establish connectivity and storage capabilities. These devices vary in specifications, such as computation power, power supply, and reporting capabilities. For instance, smart meters are capable of executing complex computations, while smart bulbs are limited to simpler tasks. Most devices within the things layer are resource-constrained and operate under limited energy, making them unsuitable for resource-intensive operations [47].

The cloud layer represents a robust layer with abundant resources that can support intricate computing tasks. It can extract information from extensive data storage and perform advanced computations, such as distributed intrusion detection. Conversely, the edge layer, also referred to as the gateway or fog layer, acts as a bridge between the resource-limited things layer and the resource-rich cloud layer. The edge layer holds significant importance within the IoT architecture, as devices in this layer are directly connected to physical things or are only a few hops away. Edge devices often possess ample resources, including storage, power supply, and computing power [53].

Each layer within the IoT architecture possesses distinct characteristics that render it indispensable. Proper organization and collaboration among these layers are crucial for building a cohesive and efficient IoT system.

IoT Security Challenges

The swift propagation of IoT devices has introduced numerous security challenges. IoT devices are recurrently resource-constrained, lacking robust security measures, and are prone to various vulnerabilities. Some of the crucial security challenges in IoT systems include the following:

(a) ***Device Vulnerabilities:*** IoT devices may have weak authentication mechanisms, outdated firmware, or insecure communication protocols, making them susceptible to attacks [31].
(b) ***Data Integrity and Privacy:*** Ensuring the integrity and privacy of data generated and communicated by IoT devices is crucial. Unauthorized access, data tampering, and privacy breaches pose significant risks [15].
(c) ***Scalability:*** The sheer number of IoT devices in large-scale deployments makes it challenging to manage security across the network effectively [36].
(d) ***Interoperability:*** Different IoT devices and platforms often use different communication protocols and data formats, leading to interoperability challenges and potential security gaps [50].
(e) ***Distributed Denial of Service (DDoS) Attacks:*** IoT devices can be conceded and used as part of botnets to introduce DDoS attacks, causing service disruptions and affecting system availability [25].

2.3 The Role of Blockchain in IoT Security

Blockchain technology offers several potential advantages in addressing IoT security challenges:

(a) ***Data Integrity and Immutability:*** The tamper-resistant nature of blockchain warrants the integrity and immutability of data. Transactions logged on the blockchain cannot be altered, providing a trustworthy source of information [54].
(b) ***Decentralized Trust***: The decentralized nature of blockchain disregards the need for a central authority, plummeting the risk of single points of failure and enhancing trust among IoT devices [12].
(c) ***Secure Data Exchange:*** Blockchain facilitates secure peer-to-peer transactions and data exchange between IoT devices without the need for mediators, reducing vulnerabilities associated with centralized servers [69].
(d) ***Enhanced Identity and Access Management:*** Blockchain-based identity management systems allow secure and verifiable authentication and access control for IoT devices [10].
(e) ***Auditability and Transparency:*** The transparent nature of blockchain enables auditing and traceability of IoT transactions, making it easier to detect and investigate malicious activities [9].

Integrating blockchain technology with IoT security can provide a more robust and resilient security framework. However, challenges related to scalability, privacy, and interoperability require attention to effectively leverage blockchain technology in IoT security applications. Machine learning algorithms can complement blockchain technology by providing intelligent threat detection, anomaly detection, and pattern recognition capabilities, thereby further enhancing the security of IoT systems.

3 Machine Learning Techniques for IoT Security

3.1 An Overview of Machine Learning in IoT Security

Machine learning algorithms can be trained on large datasets to absorb patterns and identify abnormal behavior in IoT networks. They can analyze various types of data, including network traffic, device behavior, and sensor readings to detect potential threats and intrusions [33].

Supervised learning algorithms, such as Support Vector Machines (SVM), Random Forests, and Neural Networks, can be used to categorize IoT data into different classes, such as normal or malicious behavior. Unsupervised learning algorithms, such as clustering algorithms and anomaly detection techniques, can detect anomalies in IoT data without the need for labeled training data. In Fig. 7.3, the significance of machine learning in enhancing IoT security is vividly depicted.

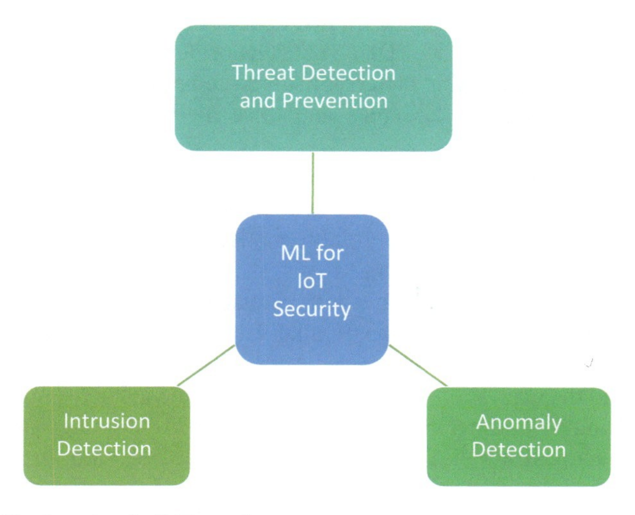

Fig. 7.3 Machine learning for IoT security

3.2 Machine Learning-Based Threat Detection and Prevention

Machine learning algorithms can analyze network traffic patterns to identify potential threats and prevent attacks in real time. By training on historical data, these algorithms can learn to recognize known attack patterns and flag distrustful activities. Intrusion Detection Systems (IDS) and Intrusion Prevention Systems (IPS) can be enhanced using machine learning techniques to identify and mitigate attacks targeting IoT systems [15].

3.3 Anomaly Detection in IoT Networks Using Machine Learning

Anomaly detection methods are employed to identify unusual behavior or outliers in IoT networks. Machine learning algorithms, such as One-Class SVM, Autoencoders, and Isolation Forests, can learn the normal behavior of IoT devices and detect deviations from the expected patterns. This approach enables the detection of previously unknown attacks or abnormal activities that may not fit predefined rules or signatures [56].

3.4 Machine Learning for Intrusion Detection in IoT

Machine learning algorithms can be trained to detect specific types of attacks, such as Distributed Denial of Service (DDoS) attacks or device spoofing. By analyzing network traffic, communication patterns, and device behavior, these algorithms can identify signs of malicious activities and raise alerts or take preventive measures [42].

Ensemble learning techniques, which combine several machine learning models, can enhance the accuracy and robustness of IoT security systems. By leveraging the strengths of different algorithms, ensemble models can enhance the detection and prevention of various types of attacks [58].

It is vital to note that the effectiveness of machine learning techniques in IoT security depends on the quality of the training data, feature selection, and the ability to adapt to developing attack techniques. Additionally, the resource constraints of IoT devices and the need for real-time processing pose challenges in implementing machine learning algorithms in IoT security systems. Addressing these challenges is crucial to harness the full potential of machine learning in securing IoT environments.

4 Integration of Blockchain Technology with Machine Learning

The integration of blockchain technology with machine learning forms a compelling synergy that holds promise for revolutionizing various domains. By combining the inherent security and decentralization of blockchain with the intelligence and adaptability of machine learning, innovative solutions are emerging to address intricate challenges across fields such as IoT security, data privacy, and transparent decision-making. Figure 7.4 illustrates the various purposes for integrating blockchain with machine learning techniques.

4.1 Data Privacy and Security

A significant benefit of integrating blockchain technology with machine learning is the enhanced data privacy and security it delivers. Blockchain's inherent features, such as decentralized storage, encryption, and immutability, can safeguard sensitive data used in machine learning models. By storing data on the blockchain, machine learning algorithms can securely access and train models without compromising the privacy and integrity of the underlying data [45].

4.2 Federated Learning on the Blockchain

Federated learning represents a decentralized methodology enabling machine learning models to be trained using data distributed across numerous devices or nodes, eliminating the necessity of transmitting data to a central server. By integrating federated learning with blockchain technology, the training process can be further decentralized and made more secure. Each participant in the network can maintain

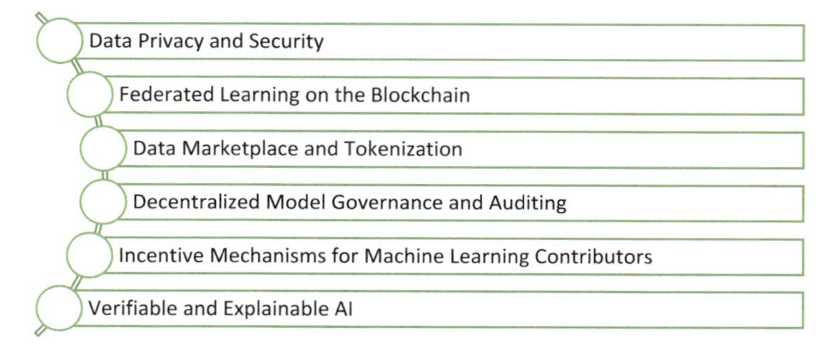

Fig. 7.4 Blockchain integration with machine learning

control over their data while contributing to the collective learning process. The blockchain ensures the integrity of the model updates and provides a transparent record of the training process [67].

4.3 Data Marketplace and Tokenization

Blockchain technology can enable secure and transparent data marketplaces where individuals or organizations can sell their data for machine learning purposes. By tokenizing data and using smart contracts, data providers can maintain control over their data and ensure fair compensation. Machine learning algorithms can access these data marketplaces, leveraging a diverse range of datasets for training models. The blockchain provides an auditable record of data transactions and ensures the integrity of the data used in the training process [19].

4.4 Decentralized Model Governance and Auditing

Integrating blockchain technology with machine learning allows for decentralized model governance and auditing. Smart contracts on the blockchain can govern the access, usage, and updates of machine learning models. This decentralized governance ensures transparency and accountability in the model development and deployment process. Additionally, the blockchain's immutability allows for independent auditing of the models, ensuring compliance and fairness [41].

4.5 Incentive Mechanisms for Machine Learning Contributors

Blockchain-based incentive mechanisms, such as token rewards or cryptocurrency payments, can be employed to incentivize individuals or organizations to contribute their computational resources or data for machine learning tasks. This incentivization promotes participation and collaboration in machine learning projects, leading to the expansion of more robust and accurate models [59].

4.6 Verifiable and Explainable AI

Blockchain technology can provide a framework for confirming the verifiability and explainability of machine learning models. The transparent and immutable nature of the blockchain allows stakeholders to trace the decisions and predictions made by machine learning models back to the input data and training process. This verifiability enhances trust and accountability in the deployment of AI systems [24].

The integration of blockchain technology with machine learning holds great potential for enhancing data privacy, security, and transparency in machine learning applications. It enables decentralized and secure training processes, fair data marketplaces, auditable model governance, and incentivized collaborations. By leveraging the strengths of both technologies, the integration of blockchain and machine learning can address critical challenges and foster the development of trustworthy and robust AI systems.

5 Current Developments in Machine Learning Approaches for Blockchain-Based IoT Security

5.1 Case Studies and Use Cases

Recent developments in machine learning approaches for blockchain-based IoT security have witnessed the emergence of various case studies and use cases. These real-world implementations showcase the effectiveness of combining machine learning and blockchain technologies to enhance IoT security. For example, machine learning algorithms have been utilized for anomaly detection in smart home systems, identifying unusual behavior patterns that could indicate potential security breaches [48]. Additionally, machine learning-based intrusion detection systems integrated with blockchain technology have been deployed in industrial IoT environments, enabling real-time threat detection and response [22].

Adaptive Threat Detection Current developments in machine learning for blockchain-based IoT security focus on adaptive threat detection techniques. These techniques involve training machine learning models on real-time IoT data to detect and respond to emerging threats effectively. By continuously updating the models with new data, adaptive threat detection algorithms can adapt to evolving attack techniques and improve the accuracy of threat identification [27].

Privacy-Preserving Machine Learning Privacy is a critical concern in IoT systems, and recent developments in machine learning aim to address this challenge. Privacy-preserving machine learning techniques, such as secure multiparty computation, homomorphic encryption, and federated learning, enable IoT devices to collaborate in training models without exposing sensitive data. These techniques ensure data privacy and confidentiality while maintaining the utility of machine learning algorithms [68].

Explainable AI for IoT Security Explainable AI (XAI) techniques are gaining attention in the context of blockchain-based IoT security. XAI focuses on providing transparent and understandable explanations for the conclusions made by machine learning models. By understanding the rationale behind the model's predictions or decisions, stakeholders can gain insights into potential vulnerabilities and mitigate security risks effectively [23].

Transfer Learning and Model Generalization Transfer learning, a technique that uses knowledge gained from one domain to boost performance in another, is being explored for enhancing the security of blockchain-based IoT systems. By training machine learning models on large-scale datasets from different IoT domains, transfer learning enables the generalization of knowledge and the detection of common security threats across diverse IoT deployments [62].

Blockchain-Based Intrusion Detection Systems Blockchain technology is being integrated into the design of intrusion detection systems (IDS) for IoT security. By using the blockchain as a tamper-resistant and transparent ledger, IDS can securely record and analyze network traffic and device behavior. Machine learning algorithms can be trained on blockchain-stored data to detect anomalies and identify potential attacks in IoT networks [1].

Reinforcement Learning for IoT Security Reinforcement learning, a branch of machine learning concerned with learning optimal decision-making strategies, is being explored for IoT security. By modeling the IoT security environment as a reinforcement learning problem, agents can learn to make adaptive security decisions based on real-time feedback. Reinforcement learning techniques can enhance the resilience and response capabilities of IoT security systems [49].

Blockchain-Enabled Trust and Reputation Systems Machine learning algorithms are being used in combination with blockchain technology to develop trust and reputation systems for IoT security. These systems leverage machine learning techniques to analyze and assess the trustworthiness of IoT devices and participants in the blockchain network. By assigning reputation scores and evaluating past behavior, these systems enhance the security and reliability of IoT transactions and interactions [18].

These current developments in machine learning approaches for blockchain-based IoT security demonstrate the ongoing research efforts to address the unique security challenges in IoT environments. By combining the strengths of machine learning algorithms with the tamper-resistant and decentralized nature of blockchain technology, these approaches offer promising solutions for enhancing the security and resilience of IoT systems. However, further research is needed to optimize these techniques, address scalability concerns, and ensure practical implementations in real-world IoT deployments.

5.2 Integration Challenges and Practical Considerations

While the integration of machine learning approaches with blockchain-based IoT security shows great potential, there are several challenges and practical considerations that are essential to be addressed:

Data Availability and Quality Machine learning models involve large volumes of high-quality data for effective training. However, in IoT environments, data availability and quality can be challenging due to the decentralized nature of the network and the limitations of IoT devices. Future research should focus on developing techniques to ensure sufficient and reliable data availability for training machine learning models in blockchain-based IoT security [7].

Computational Resources Machine learning algorithms are computationally intensive, and deploying them on resource-constrained IoT devices can be a challenge. Considerations should be given to optimizing the computational resources required for running machine learning algorithms on IoT devices and leveraging distributed computing architectures to alleviate the computational burden [34].

Latency and Real-Time Processing IoT systems often require real-time or near real-time processing to detect and retort to security threats promptly. Integrating machine learning with blockchain technology should account for the potential latency introduced by the consensus mechanisms of the blockchain. Efficient techniques should be developed to enable real-time processing and decision-making in blockchain-based IoT security [35].

Privacy and Compliance IoT systems often handle sensitive data, such as personal or healthcare information, which necessitates adherence to privacy regulations. Integrating machine learning approaches with blockchain technology should prioritize privacy-preserving techniques to confirm compliance with data protection regulations. Mechanisms such as federated learning or secure multi-party computation can be explored to maintain data privacy while leveraging the benefits of machine learning [38].

Scalability and Blockchain Throughput Scalability issues emerge in blockchain networks when confronted with a substantial volume of transactions, particularly evident within IoT environments encompassing a multitude of devices. To integrate machine learning with blockchain-based IoT security, scalable blockchain solutions and optimized consensus mechanisms should be explored to handle the increasing throughput requirements [44].

6 Open Challenges and Future Directions

6.1 Scalability

Enhancing the scalability remains a noteworthy challenge in the integration of machine learning approaches with blockchain-based IoT security. As the number of IoT devices and transactions rises, the computational and storage necessities for training and maintaining machine learning models on the blockchain become more demanding. Future research could focus on the following methodologies:

- Conduct a comparative study of existing consensus mechanisms, such as Proof of Stake (PoS), Proof of Authority (PoA), and sharding, to assess their suitability for handling increased transaction throughput while maintaining security [16].
- Develop a simulation framework to assess the performance of various scalability solutions under different network loads and transaction rates. Additionally, investigate off-chain solutions like state channels and sidechains to alleviate scalability limitations and enhance overall network efficiency [26].

6.2 Privacy-Preserving Machine Learning

While privacy-preserving machine learning techniques have made significant advancements, there is still a need for robust methods to ensure data privacy in blockchain-based IoT security while leveraging the benefits of machine learning. Developing efficient and practical privacy-preserving mechanisms that can guard sensitive data while maintaining the utility of machine learning models remains an open challenge.

- Implement federated learning to train machine learning models on decentralized data sources without sharing raw data [66].
- Investigate progressive cryptographic techniques like homomorphic encryption and secure multiparty computation to ensure data privacy during model aggregation [17].
- Develop a framework to evaluate the trade-off between privacy preservation and model accuracy using real-world IoT datasets. Explore techniques to control and manage the privacy budget in federated learning [43].

6.3 Interoperability and Standardization

Interoperability and standardization are crucial for the extensive adoption of blockchain-based IoT security solutions. A potential research problem in this context is how we can establish seamless communication among diverse IoT devices using different communication protocols and data formats. As different blockchain platforms and IoT devices use varying protocols and data formats, interoperability challenges arise. Future research should focus on the following:

- Develop a protocol translation mechanism that enables devices using different communication protocols to communicate effectively [46].
- Investigate the adoption of standardized data formats such as the SensorThings API to promote interoperability [28].
- Implement the Web of Things (WoT) ontology to provide a standardized framework for data exchange and semantic interoperability [14] and conduct case studies to assess the effectiveness of these approaches in real-world IoT environments.

6.4 Adversarial Attacks and Defense

Detection and mitigation of adversarial attacks targeting machine learning models pose a significant threat to the security of blockchain-based IoT systems. Adversaries can manipulate input data or inject malicious behavior to deceive the machine learning algorithms. Mechanisms such as adversarial training and detection enables us to develop robust defense mechanism and mitigating these attacks is an important area of future research.

- Develop machine learning-based intrusion detection systems that specialize in identifying adversarial attacks [2].
- Implement adversarial training, where models are trained on adversarial examples to enhance their robustness [8].
- Investigate novel anomaly detection techniques, such as adversarial anomaly detection, to identify adversarial patterns that deviate from normal behavior [61] and evaluate the performance of these approaches using benchmark datasets and consider real-world deployment scenarios.

6.5 Energy Efficiency

Energy efficiency is a critical consideration in resource-constrained IoT environments. Machine learning algorithms, especially complex deep learning models, can consume significant computational resources and energy. Future research should concentrate on developing energy-efficient machine learning techniques tailored for IoT devices, enabling efficient utilization of resources without compromising security. The possible methodologies to reduce computational burden of machine learning on resource constrained IoT devices should be reduced to achieve energy efficiency are as follows:

- Explore edge-based processing, where machine learning models are deployed on devices with sufficient resources, reducing the need for extensive data transmission [60].
- Implement model compression techniques like quantization and pruning to reduce the computational complexity of models [30] and develop a framework to assess the trade-off between model accuracy and computational efficiency in energy-constrained IoT scenarios.

6.6 Governance and Trust

The governance and trust mechanisms for machine learning models in blockchain-based IoT security need further exploration. Developing decentralized governance models that ensure transparency, accountability, and fairness in the deployment and management of machine learning models is crucial. Trust frameworks that establish the reliability and integrity of machine learning models and their interactions within the blockchain network are important future directions.

- Implement smart contracts to establish transparent rules and automate trust-building processes [55].
- Develop reputation systems using blockchain technology to assess the trustworthiness of participants based on historical behavior [21].
- Integrate blockchain with digital identity systems to enhance user and device authentication and establish a trustworthy identity framework [32]. Conduct simulations or case studies to evaluate the effectiveness of these mechanisms in building trust within IoT networks.

6.7 Real-Time Adaptability

Enabling real-time adaptability of machine learning models in blockchain-based IoT security for rapidly evolving threats is a challenge. IoT systems require immediate response and adaptation to dynamic security threats. Future research should emphasize on developing efficient algorithms and architectures that can enable real-time learning, decision-making, and model updates in IoT environments.

- Investigate online learning techniques that enable machine learning models to adapt continuously to new data and threats [3].
- Develop a dynamic update mechanism that incorporates new threat information into existing models in real time [64].
- Implement reinforcement learning approaches to enable agents to make adaptive security decisions based on real-time feedback [65] and evaluate the responsiveness and effectiveness of these methods using real-world IoT security scenarios.

6.8 Ethical and Legal Considerations

As machine learning techniques and blockchain-based IoT security solutions evolve, ethical and legal considerations need to be addressed. Ensuring fairness, transparency, and accountability in the collection, processing, and use of data is essential. Future research should explore ethical frameworks and regulatory guidelines to govern the incorporation of machine learning and blockchain technologies in IoT security.

- Conduct a comprehensive review of existing ethical and legal frameworks relevant to IoT and machine learning security [29].
- Collaborate with legal experts to develop guidelines for data handling, consent, and user rights in the context of IoT security [57].
- Create a decision-making framework that balances security requirements with ethical considerations and validate it through case studies or simulations [39].

Tackling these existing challenges and venturing into future avenues will establish a path toward creating resilient, scalable, and privacy-centric machine learning strategies for ensuring the security of IoT systems integrated with blockchain technology. By advancing the state of the art in these areas, researchers and practitioners can create a more secure and trustworthy environment for the deployment and operation of IoT systems.

7 Conclusion

7.1 Key Findings

In this chapter, we have delved into the integration of machine learning approaches with blockchain technology in the context of IoT security. We have examined the current developments, identified open challenges, and discussed future directions in this rapidly evolving field. Here are the key findings:

(i) Machine learning techniques, such as anomaly detection, threat detection, and pattern recognition, enhance the security of IoT systems by detecting and mitigating potential cyber threats.
(ii) Blockchain technology provides decentralized storage, immutability, and transparent transactions, ensuring data integrity and privacy in IoT environments.
(iii) Privacy-preserving machine learning techniques, federated learning, and explainable AI are crucial for addressing privacy concerns and ensuring transparency and interpretability in machine learning models.

(iv) Transfer learning, reinforcement learning, and blockchain-enabled trust mechanisms hold promise for improving the resilience and generalizability of machine learning approaches in IoT security.
(v) Scalability, privacy preservation, interoperability, and defense against adversarial attacks are major challenges that need to be overcome for effective integration of machine learning and blockchain in IoT security.

7.2 Recommendations for Future Research

Derived from the findings, here are several suggestions for prospective research endeavors:

(i) Develop scalable machine learning algorithms and distributed architectures that can manage the escalating volume and complexity of IoT data in blockchain-based systems.
(ii) Further explore privacy-preserving techniques that allow secure collaboration and training of machine learning models while preserving data privacy and confidentiality.
(iii) Standardize protocols and frameworks to enable interoperability between different blockchain platforms and IoT devices, promoting seamless integration and communication.
(iv) Enhance defense mechanisms against adversarial attacks by developing robust techniques such as adversarial training, detection, and robust model architectures.
(v) Focus on energy-efficient machine learning approaches tailored for resource-constrained IoT devices to optimize energy consumption and prolong device lifespan.
(vi) Investigate decentralized governance models and trust frameworks to ensure transparency, accountability, and fairness in the deployment and management of machine learning models in blockchain-based IoT security.
(vii) Address real-time adaptability requirements by developing algorithms and architectures that enable continuous learning, decision-making, and model updates in dynamic IoT environments.
(viii) Establish ethical frameworks and regulatory guidelines to govern the ethical and legal considerations surrounding the collection, processing, and use of data in machine learning and blockchain-based IoT security.

By addressing these recommendations and further exploring the intersection of machine learning and blockchain technology in IoT security, researchers can make significant strides in creating robust, scalable, and privacy-preserving solutions. This will contribute to the development of a more secure and trustworthy IoT ecosystem, benefiting various sectors such as healthcare, business, and smart cities. Collaborative efforts and interdisciplinary research are key to advancing the field and realizing the full potential of machine learning in blockchain-based IoT security.

References

1. Al-Emari, S. (2021). Intrusion detection systems using blockchain technology: A review, issues and challenges. *Computer Systems Science and Engineering, 40*, 87–112. https://doi.org/10.32604/csse.2022.017941
2. Alotaibi, A. (2023). Adversarial machine learning attacks against intrusion detection systems: A survey on strategies and defense. *Future Internet, 15*(2), 62. https://doi.org/10.3390/fi15020062
3. Amin Shahraki, M. A. (2022). A comparative study on online machine learning techniques for network traffic streams analysis. *Computer Networks, 207*(22), 108836. https://doi.org/10.1016/j.comnet.2022.108836
4. Atzori, L. (2010). The internet of things: A survey. *Computer Networks, 54*, 2787. https://doi.org/10.1016/j.comnet.2010.05.010
5. Ayan Chatterjee, B. S. (2022). IoT anomaly detection methods and applications: A survey. *Internet of Things, 19*, 100568. https://doi.org/10.1016/j.iot.2022.100568
6. Buterin, V. (2017). *Ethereum: A next generation smart contract and decentralized application platform (2013)*. Retrieved from http://ethereum.org/ethereum.html
7. Byabazaire, J. G. (2020). Data quality and trust: Review of challenges and opportunities for data sharing in IoT. *Electronics, 9*(12), 2083. https://doi.org/10.3390/electronics9122083
8. Chen, H. (2022). Adversarial training for improving model robustness? Look at both prediction and interpretation. *In Proceedings of the AAAI Conference on Artificial Intelligence, 36*(10), 10463–10472.
9. Christidis, K. (2016). Blockchains and smart contracts for the internet of things. *IEEE Access, 4*, 1–1. https://doi.org/10.1109/ACCESS.2016.2566339
10. Dagher, G. (2018). Ancile: Privacy-preserving framework for access control and interoperability of electronic health records using blockchain technology. *Sustainable Cities and Society., 39*, 283. https://doi.org/10.1016/j.scs.2018.02.014
11. Dhillon, V. (2022). Blockchain enabled applications understand the blockchain ecosystem and how to make it work for you.
12. Dorri, A. K. (2016). Blockchain in internet of things: Challenges and solutions. *arXiv preprint arXiv:1608.05187.*
13. Dorri, A. K. (2017). *Blockchain for IoT security and privacy: The case study of a smart home*. https://doi.org/10.1109/PERCOMW.2017.7917634
14. Antoniazzi, F., & Viola, F. (2019). Building the semantic web of things through a dynamic ontology. *IEEE Internet of Things Journal, 6*(6), 10560–10579. https://doi.org/10.1109/JIOT.2019.2939882
15. Fadele, A. (2017). Internet of things security: A survey. *Journal of Network and Computer Applications., 88*, 10. https://doi.org/10.1016/j.jnca.2017.04.002
16. Fahim, S. (2023). Blockchain: A comparative study of consensus algorithms PoW, PoS, PoA, PoV. I. *Journal of Mathematical Sciences and Computing, 3*, 46–57. https://doi.org/10.5815/ijmsc.2023.03.04
17. Fang, H., & Q. Q. (2021). Privacy preserving machine learning with homomorphic encryption and federated learning. *Future Internet, 13*(4), 94. https://doi.org/10.3390/fi13040094
18. Putrat, G. D., & S. M. (2023). *Trust and reputation management for blockchain-enabled IoT* (pp. 529–536). 2023 15th International Conference on COMmunication Systems & NETworkS (COMSNETS), Bangalore, India. https://doi.org/10.1109/COMSNETS56262.2023.10041348
19. Zyskind, G., & O. N. (2015). *Decentralizing privacy: Using blockchain to protect personal data* (pp. 180–184). IEEE Security and Privacy Workshops. https://doi.org/10.1109/SPW.2015.27
20. Gubbi, J. (2012). Internet of things (IoT): A vision, architectural elements, and future directions. *Future Generation Computer Systems, 29*, 1645. https://doi.org/10.1016/j.future.2013.01.010
21. Hemmrich, S. (2023). Business Reputation Systems Based on Blockchain Technology—A Risky Advance

22. Islam, N. (2021). Towards machine learning based intrusion detection in IoT networks. *Cmc –Tech Science Press, 69*, 1801–1821. https://doi.org/10.32604/cmc.2021.018466
23. Jagatheesaperumal, S. K. (2022). Explainable AI over the internet of things (IoT): Overview, state-of-the-art and future directions. *IEEE Open Journal of the Communications Society, 3*, 2106.
24. Javed, A., Ahmed, W., Pandya, S., Maddikunta, P., Alazab, M., & Gadekallu, T. (2023). A survey of explainable artificial intelligence for smart cities. *Electronics, 12*, 1020. https://doi.org/10.3390/electronics12041020
25. Khatkar, M. (2020). *An overview of distributed denial of service and internet of things in healthcare devices* (pp. 44–48). https://doi.org/10.1109/INBUSH46973.2020.9392171.
26. Kim, S. (2018). A survey of scalability solutions on blockchain. *Conference: 2018 International Conference on Information and Communication Technology Convergence (ICTC)*, (pp. 1204–1207). https://doi.org/10.1109/ICTC.2018.8539529
27. Kiran, A., Mathivanan, P., Mahdal, M., Sairam, K., Chauhan, D., & Talasila, V. (2023). Enhancing data security in IoT networks with blockchain-based management and adaptive clustering techniques. *Mathematics, 11*, 2073. https://doi.org/10.3390/math11092073
28. Kotsev, A., & S. K. (2018). Extending INSPIRE to the internet of things through SensorThings API. *Geosciences, 8*(6), 221. https://doi.org/10.3390/geosciences8060221
29. Lee. (2020). Internet of things (IoT) cybersecurity: Literature review and IoT cyber risk management. *Future Internet, 12*(9), 157. https://doi.org/10.3390/fi12090157
30. Li, Z. H. (2023). Model compression for deep neural networks: A survey. *Computers, 12*(3), 60. https://doi.org/10.3390/computers12030060
31. López Vargas, A. (2020). Challenges and opportunities of the internet of things for global development to achieve the United Nations sustainable development goals. *IEEE Access, 1-1*, 37202. https://doi.org/10.1109/ACCESS.2020.2975472
32. Lukas Stockburger, G. K. (2021). Blockchain-enabled decentralized identity management: The case of self-sovereign identity in public transportation. *Blockchain: Research and Applications, 2*(2), 100014. https://doi.org/10.1016/j.bcra.2021.100014
33. Al-Garadi, M. A., & A. M.-A. (2020). A survey of machine and deep learning methods for internet of things (IoT) security. *IEEE Communications Surveys & Tutorials, 22*(3), 1646–1685. https://doi.org/10.1109/COMST.2020.2988293
34. Li, M., & F. R. (2022). Intelligent resource optimization for blockchain-enabled IoT in 6G via collective reinforcement learning. *IEEE Network, 36*(6), 175–182. https://doi.org/10.1109/MNET.105.2100516
35. Khan, M. A., & K. S. (2018). IoT security: Review, blockchain solutions, and open challenges. *Future Generation Computer Systems, 82*, 395–411. https://doi.org/10.1016/j.future.2017.11.022
36. Miorandi, D. (2012). Internet of things: Vision, applications and research challenges. *Ad Hoc Networks, 10*, 1497. https://doi.org/10.1016/j.adhoc.2012.02.016
37. Mizrahi, I. B. (2016). Cryptocurrencies without proof of. *Work, 9604*, 142–157. https://doi.org/10.1007/978-3-662-53357-4_10
38. Mohanta, B. (2020). Addressing security and privacy issues of IoT using Blockchain technology. *IEEE Internet of Things Journal, 8*, 1–1. https://doi.org/10.1109/JIOT.2020.3008906
39. Mökander, J. M. (2021). Ethics-based auditing of automated decision-making systems: Nature, scope, and limitations. *Science and Engineering Ethics, 27*, 44. https://doi.org/10.1007/s11948-021-00319-4
40. Nakamoto, S. (2009). *Bitcoin: A peer-to-peer electronic cash system*. Cryptography. Mailing list at https://metzdowd.com
41. Nassar, M. (2020). *Blockchain for explainable and trustworthy artificial intelligence* (Vol. 10, p. 10). Wiley Interdisciplinary Reviews: Data Mining and Knowledge Discovery. https://doi.org/10.1002/widm.1340

42. Nene, M. (2013). A survey on machine learning techniques for intrusion detection systems. *International Journal of Advanced Research in Computer and Communication Engineering, 11*, 4349.
43. Nguyen Truong, K. S. (2021). Privacy preservation in federated learning: An insightful survey from the GDPR perspective. *Computers & Security, 110*, 102402. https://doi.org/10.1016/j.cose.2021.102402
44. Novo, O. (2019). Scalable access management in IoT using blockchain: A performance evaluation. *IEEE Internet of Things Journal, 6*, 4694–4701. https://doi.org/10.1109/JIOT.2018.2879679
45. Ouaddah, A. (2017). FairAccess: A new Blockchain-based access control framework for the internet of things: FairAccess: A new access control framework for IoT. *Security and Communication Networks, 9*, 5943. https://doi.org/10.1002/sec.1748
46. Sethil, P., & Sarangi, S. R. (2017). Internet of things: Architectures, protocols, and applications. *Journal of Electrical and Computer Engineering, 2017*, 1. https://doi.org/10.1155/2017/9324035
47. Perera, C. (2015). The emerging internet of things marketplace from an industrial perspective: A survey. *IEEE Transactions on Emerging Topics in Computing, 3*, 585. https://doi.org/10.1109/TETC.2015.2390034
48. Ramapatruni, S. (2019). *Anomaly detection models for smart home security* (pp. 19–24). https://doi.org/10.1109/BigDataSecurity-HPSC-IDS.2019.00015.
49. Rawat, A. U. (2021). Reinforcement learning for {IoT} security: A comprehensive survey. *IEEE – Internet of Things Journal, 8*(11), 8693–8706. https://doi.org/10.1109/jiot.2020.3040957
50. Roman, R. (2013). On the features and challenges of security and privacy in distributed internet of things. *Computer Networks, 57*, 2266–2279. https://doi.org/10.1016/j.comnet.2012.12.018
51. Shancang Li, L. D. (2018). 5G internet of things: A survey. *Journal of Industrial Information Integration, 10*, 1–9. https://doi.org/10.1016/j.jii.2018.01.005
52. Shi, W. (2016). Edge computing: Vision and challenges. *IEEE Internet of Things Journal, 3*, 1–1. https://doi.org/10.1109/JIOT.2016.2579198
53. Stojmenovic, M. W. (2014). Introduction to the special issue: Fog computing in future generation communication networks. *Journal of Network and Computer Applications, 45*, 1–3.
54. Swan, M. (2015). *Blockchain: Blueprint for a new economy*. O'Reilly Media.
55. Taherdoost, H. (2023). Smart contracts in Blockchain technology: A critical review. *Information, 14*(2), 117. https://doi.org/10.3390/info14020117
56. Tahsien, S. (2020). Machine learning based solutions for security of internet of things (IoT): A survey. *Journal of Network and Computer Applications, 161*, 102630. https://doi.org/10.1016/j.jnca.2020.102630
57. Tawalbeh, L. F. (2020). IoT privacy and security: Challenges and solutions. *Applied Sciences, 10*(12), 4102. https://doi.org/10.3390/app10124102
58. Tomer, V., & Sharma, S. (2022). Detecting IoT attacks using an ensemble machine learning model. *Future Internet, 14*, 102. https://doi.org/10.3390/fi14040102
59. Toyoda, K. (2019). *Mechanism design for an incentive-aware blockchain-enabled federated learning platform* (pp. 395–403). https://doi.org/10.1109/BigData47090.2019.9006344.
60. Truong, H. T.-H. (2023). Making distributed edge machine learning for resource-constrained communities and environments smarter: Contexts and challenges. *Journal of Reliable Intelligent Environments, 9*, 119–134. https://doi.org/10.1007/s40860-022-00176-3
61. Xianchao Zhang, J. M. (2022). Deep anomaly detection with self-supervised learning and adversarial training. *Pattern Recognition, 121*, 108234. https://doi.org/10.1016/j.patcog.2021.108234
62. Xiaoding, W. (2021). Enabling secure authentication in industrial IoT with transfer learning empowered blockchain. *IEEE Transactions on Industrial Informatics, 17*, 1–1. https://doi.org/10.1109/TII.2021.3049405

63. Yang, F. (2019). Delegated proof of stake with downgrade: A secure and efficient blockchain consensus algorithm with downgrade mechanism. *IEEE Access, 7*, 1–1. https://doi.org/10.1109/ACCESS.2019.2935149
64. Yu Zheng, Z. L. (2022). Dynamic defenses in cyber security: Techniques, methods and challenges. *Digital Communications and Networks, 8*(4), 422–435. https://doi.org/10.1016/j.dcan.2021.07.006
65. Yunhan Huang, L. H. (2022). Reinforcement learning for feedback-enabled cyber resilience. *Annual Reviews in Control, 53*, 273–295. https://doi.org/10.1016/j.arcontrol.2022.01.001
66. Zhang, C. (2021). A survey on federated learning. *Knowledge-Based Systems., 216*, 106775. https://doi.org/10.1016/j.knosys.2021.106775
67. Zhao, Y. (2020). Privacy-preserving Blockchain-based federated learning for IoT devices. *IEEE Internet of Things Journal, 8*, 1–1. https://doi.org/10.1109/JIOT.2020.3017377
68. Zheng, M. X. (2019). *Challenges of privacy-preserving machine learning in IoT* (pp. 1–7). In Proceedings of the First International Workshop on Challenges in Artificial Intelligence and Machine Learning for Internet of Things.
69. Zheng, Z., & N. (2017). An overview of blockchain technology: Architecture. *Consensus, and Future Trends.* https://doi.org/10.1109/BigDataCongress.2017.85

Chapter 8
Decentralized Identity Management Using Blockchain Technology: Challenges and Solutions

Ahmed Mateen Buttar, Muhammad Anwar Shahid, Muhammad Nouman Arshad, and Muhammad Azeem Akbar

1 Introduction

Blockchain-powered decentralized identity management provides consumers with authority over their personal data and increases online company privacy, security, and interoperability. Traditional identity systems are administered by governments or enterprises. Blockchain, on the other hand, empowers users in its decentralized identity management. Blockchain records transactions and data across computers. Cryptographically linking each transaction, or "block," creates an immutable chain of information. Decentralized and tamper-resistant blockchain is ideal for secure and transparent identity management [1].

Decentralized identity management systems create and manage digital IDs by using cryptography. Blockchain secures their identities. Selectively sharing names, ages, and addresses with numerous service providers avoids providing unnecessary personal information. Decentralized identity management leads to self-sovereignty. Individuals can manage their identities without intermediaries. Controlling data access prevents identity theft, fraud, and misuse. Interoperability improves with decentralized identity management. The Decentralized Identity Foundation (DIF) and World Wide Web Consortium (W3C) verifiable credentials (VCs) make verifying and sharing digital identities across platforms, apps, and organizations easy [2].

Blockchain secures identity management. Cryptography secures blockchain transactions and identities. Blockchain's decentralized nature makes it more secure.

A. M. Buttar (✉) · M. N. Arshad
Department of Computer Science, University of Agriculture Faisalabad, Faisalabad, Pakistan

M. A. Shahid
Univeristy of Windsor, Windsor, ON, Canada

M. A. Akbar
Department of Software Engineering, LUT University, Lappeenranta, Finland

S. M. Idrees, M. Nowostawski (eds.), *Blockchain Transformations*, Signals and Communication Technology, https://doi.org/10.1007/978-3-031-49593-9_8

Blockchain platforms are developing decentralized identity management. Sovrin, uPort, and Microsoft's ION leverage blockchain develop open, interoperable, and privacy-preserving identity systems for computer-to-computer transactions.

1.1 Background and Motivation

Traditional centralized identification systems' flaws prompted blockchain-based decentralized identity management. Centralized identification systems compel users to share sensitive personal data with several service providers, increasing the risk of data breaches and identity theft. Hackers target centralized databases, which can harm individuals. Decentralized identity management enables people to own their data and share only what they need to.

Less user control: Traditional identification systems govern identities centrally. Unauthorized data collection, storage, and use occur frequently. Decentralized identity management empowers individuals. Users can choose who sees what and why. Centralized identity schemes are also incompatible. Each service provider or organization may have its own identity databases and verification methods, causing duplications in data collection and poor user experiences. Decentralized identity management using open standards and protocols lets people use verifiable credentials across platforms and services, increasing interoperability [5].

Identity fragmentation: People have multiple digital identities across platforms and services. Password management is challenging. Decentralized identity management solves this identity fragmentation by providing a single, portable, self-sovereign identity that can be used across apps and services.

Credibility: Identity management demands trust. Traditional systems verify IDs centrally. These authorities may be corrupt or make mistakes. Decentralized identity management verifies credentials by using blockchain's immutability and transparency. Cryptographically linking IDs to the blockchain prevents forgery.

1.2 Objectives and Scope

Blockchain-based decentralized identity management seeks user control. Digital identities must be controlled to create, manage, and cancel identities. They can also select who can access their personal data and how. Decentralized identity management emphasizes privacy and security. Personal data should be protected, and cryptography and blockchain's immutability safeguard identities and attributes against manipulation and fraud. The interoperability of systems, services, and organizations are essential. Decentralized identity management should streamline the exchange and verification of digital identities and credentials, improving user experiences. This goal requires protocol and format standardization.

Digital identities must be trusted. Blockchain's immutability should offer a tamper-proof way to verify identification credentials. Cryptographic and consensus algorithms maintain system confidence. Decentralized identity management solutions must be usable and adopted by users. For this, simple and easy identity management is needed. Developers, stakeholders, and service providers must work together to create user-friendly apps and interfaces. The decentralized identity management system must be able to handle multiple users and transactions. Digital identities and transactions should keep the system efficient and responsive. Decentralized identity management must comply with regulations and laws. The system must preserve data and privacy. Identity issues and legal recourse should be addressed [6].

1.3 Overview of Blockchain Technology

Blockchain-based decentralized identity management is secure, transparent, and user centric. Blockchain identity management boasts many features.

Distributed ledgers: Blockchain stores identity-related data in a decentralized, tamper-resistant ledger. Each identity-related transaction, like generating, modifying, or canceling an identity, is stored in a block and connected to the previous block, producing an immutable data chain.

Self-sovereign identities: Blockchain enables people to control their digital identities. Public–private key pairs create and manage blockchain identities, thus eliminating centralized identity validation and administration. Blockchain-identifying records are unchangeable and transparent. This secures and verifies the identification data. Blockchain transparency allows people and businesses to verify identity credentials without needing intermediaries [7].

Digital credentials: Blockchain can issue and validate digital credentials. Parties who rely on blockchain can verify issuers' public key–signed credentials. Verifying identification permits selective sharing. Standards enable decentralized identity management interoperability. DIF and W3C open standards like decentralized identifiers (DIDs) and verifiable credentials provide blockchain-based identification system compatibility and interoperability.

Privacy-preserving features: With decentralized identity management, someone can choose to share the required traits or credentials without revealing their whole identity. Attribute-based access control safeguards personal data. Blockchain security promotes decentralized identity management. Blockchain encryption safeguards IDs. Finally, blockchain's decentralized nature resists attacks.

Chapter Objectives:

Examine blockchain-based decentralized identity management difficulties.

Help blockchain-based identity systems to scale.

Present zero-knowledge proofs and confidential transactions to address privacy and security issues in decentralized identity management.

Discuss decentralized identification system interoperability issues and provide solutions for data flow between platforms.

Analyze regulatory compliance needs and suggest ways to comply with applicable laws.

Investigate user-adoption issues and propose user-centric interfaces and educational campaigns to promote decentralized identity solutions.

Showcase successful decentralized identity management installations and case studies.

Discuss blockchain-based decentralized identity management research and development.

Chapter Organization

This chapter overviews blockchain-based decentralized identity management and the benefits associated with it. After that, a discussion of the challenges posed by distributed identity management follows. Scalability, privacy and security, interoperability, regulatory compliance, and user adoption each has its own section in this chapter. This chapter provides solutions and strategies, and it explores the specific difficulties in and complexity of each challenge. The solutions include cross-chain communication, off-chain storage, layer 2 protocols, zero-knowledge proofs, governance frameworks, user-centric interfaces, and education programs. Case studies and actual implementations from the real world provide the reader with useful insights throughout the chapter. The chapter comes to a close with research and development on a type of decentralized identity management that is based on blockchain technology, as shown in Fig. 8.1 [8].

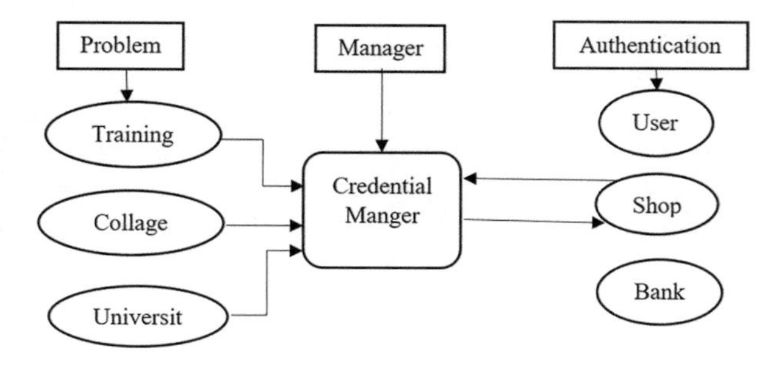

Fig. 8.1 Credential verification guide using dock certs and dock wallets

2 Fundamentals of Decentralized Identity Management

Decentralized identity management in blockchain technology has the following core concepts and components.

Decentralization: Identity management is decentralized. Instead of having a central authority control users' identities, the users control them. Users create, manage, and share their blockchain-stored digital identities. Decentralized identity management uses blockchain technology. It is a transparent, tamper-proof distributed ledger. By cryptographically linking blocks, blockchain creates an unchangeable chain of identifying data. Decentralized identity management requires self-sovereign identification. It lets people manage and confirm their identities without central authorities. Users can also create, amend, and revoke IDs. Decentralized identity management requires cryptography. It secures blockchain identification data. Public-key cryptography, digital signatures, and hash functions protect these identity data. Decentralized identity management uses digital identity traits to represent personal information. Names, ages, addresses, credentials, and other information may be included. Users can selectively share attributes and limit whom to share them with [9].

Verifiable credentials: Verifiable credentials are blockchain-issued, stored, and verified identification credentials. Reliant parties can verify them by using the issuer's public keys because they are cryptographically signed. Credentials verify identification information.

Decentralized identifiers (DIDs): Blockchain users are allocated unique identifiers. DIDs enable decentralized identity reference and interaction. They facilitate interoperability and both identity resolution and identity discovery across platforms and services.

Interoperability standards: Blockchain-based identification systems must be interoperable for decentralized identity management to work. DIDs and verifiable credentials, established by the Decentralized Identity Foundation (DIF) and the World Wide Web Consortium (W3C), enable interoperability and compatibility across platforms.

2.1 Concepts and Terminology

Blockchain-based decentralized identity management is undergirded by several concepts and terminologies.

Digital identity: A digital identity marks a person's online presence. Their name, age, address, biometric data, and other information may be included.

Decentralization: Power is distributed. Decentralized identity management lets users control their digital identities without intermediaries.

Self-sovereign identity: People can control their digital identities. Such an identity allows identity management, regulation, and authentication without third parties.

Blockchain: Blockchain stores data across computers. Cryptography secures immutable and transparent data. Blockchain underpins decentralized identity management.

Cryptography: Algorithms protect data. Decentralized identity management protects data via cryptography, public-key cryptography, digital signatures, and hash functions. Digital IDs and qualifications are available from trusted partners. Reliant parties can validate them because they are cryptographically signed by the issuer. Credentials allow selective identity disclosure.

DIDs: DIDs are unique blockchain identifiers for individuals, entities, and items. They provide decentralized identification and engagement. They aid cross-platform identity resolution and discovery. Providers issue and manage these digital IDs. Decentralized identity management utilizes people, organizations, or machines as identity providers. Digital identities can also be authenticated. Identity wallets manage digital identities and credentials. These wallets protect private keys, manage verifiable credentials, and selectively share identity information with service providers [10].

Interoperability standards: Specifications and protocols allow distributed identity management systems to communicate. Interoperability standards include DIDs, VCs, and W3C DID and VC standards. Blockchain-based decentralized identity management is based on and requires these concepts and terminology. They start user-centric, safe, and privacy-enhancing digital identity management.

2.2 Self-Sovereign Identity

Blockchain-based identity management involves self-identification. It lets people control, regulate, and authenticate their digital identities without centralized authorities. Self-sovereign identity requires user control.

User control: Self-sovereign identification allows for digital identity management. IDs can be created, edited, and revoked. Users determine identification, data access, and conditions. Users no longer need centralized identity validation or administration. Blockchain supports self-sovereign identity. Blockchain's distributed ledger protects identification data. The blockchain lets several parties verify and validate identifying information. Cryptography promotes self-sovereign identification. Public-key cryptography helps users create and control cryptographic key pairs: private keys for identity management and public keys for verification. Digitally signing and encrypting identifiers guarantees privacy, integrity, and authentication [11].

Self-sovereignty demands verifiable credentials. Trusted organizations issue digitally signed credentials that can be verified. Public keys enable the selective transmission of these credentials without revealing personal information. Self-sovereign identity systems use interoperability standards. DIDs and verifiable credentials allow cross-platform identity generation, exchange, and verification. Interoperability helps digital ecosystem elements verify identities. Self-sovereign identity applications, services, and domains to be portable. Portability eliminates the need for different accounts and credentials. Self-sovereign identification shifts confidence from authorities to users. Cryptography and blockchain transparency build trust. Cryptographic evidence, decentralized ledger integrity, and identity management build confidence. Decentralized identity management equips people with self-sovereign identification. It supports digital interoperability and privacy [12].

2.3 Decentralized Identifiers (DIDs) and Verifiable Credentials

Blockchain-based decentralized identity management requires DIDs and verifiable credentials. Digital identities and credentials can be created, exchanged, and verified in a decentralized way.

Decentralized IDs

- DIDs are assigned by decentralized identification systems. They are globally unique and platform resolvable.
- DIDs allow identity referencing and interaction without central authorities. Cryptographic key pairs and Uniform Resource Identifiers (URIs) can control and authenticate Decentralized Identifiers (DIDs).
- Blockchains and other decentralized systems can store and resolve DIDs, making identity management tamper-proof and available worldwide.

Trusted organizations issue verifiable credentials. They reveal selective identification without revealing personal information. Issuers cryptographically sign verifiable credentials. The issuer's public keys enable easy verification.

- Independently verifiable credentials include names, ages, educational qualifications, and membership statuses.
- Verifiable credentials allow users to verify, authenticate, and access services.

Verifiable credentials safeguard privacy in that users pick which credentials to reveal, when, and with whom.

DIDs and verifiable credentials support blockchain-based identity management. DIDs provide unique, decentralized DID identification and interaction, whereas verifiable credentials issue, exchange, and verify reliable, tamper-proof digital evidence of identity attributes or qualifications. These technologies provide user-centric blockchain identification, privacy, and interoperability [13].

2.4 Benefits and Use Cases

Blockchain-based decentralized identity management offers various benefits and uses. Decentralized identity management empowers and protects users. Maintaining, regulating, and selectively publishing identification attributes improves privacy and reduces centralized power.

Trust and safety: Blockchain's immutability and cryptography secure identification data. Tamper-resistant blockchains safeguard identities from fraud and data breaches. Decentralized identity management also improves platform/service compatibility. Overall, DIDs and verifiable credentials make digital life easier by helping people to recognize and trust one another.

Saving money: Decentralized identity management removes multiple service logins. Portable identities reduce administrative hassle. Decentralized identity management also lowers identity authentication costs.

2.5 Decentralized Identity Management

Digital identity verification: Decentralized identity management streamlines client onboarding, online service access, and knowing your customer (KYC). Verifiable credentials also streamline identification verification, and decentralized identity management facilitates cross-border verification. Securely communicating valid credentials with foreign authorities accelerates cross-border transactions and minimizes superfluous identification verification. Decentralized identity management lets people own their data. Allowing used to share only certain traits limits service providers' access to their personal data. Data breaches and illegal access are thereby reduced. Blockchain-based decentralized identity management can handle supply chains and product authenticity. Verifiable credentials prevent product counterfeiting and ensure supply-chain transparency. Decentralized identity management secures and exchanges patient data, improving healthcare systems. Sharing medical records with providers improves data privacy, interoperability, and care coordination. Blockchain-based identity management secures digital voting. It prevents voter fraud by verifying identities, as shown in Fig. 8.2 [14].

3 Challenges in Implementing Decentralized Identity

Blockchain-based decentralized identity management has many benefits but also many drawbacks.

Scalability: Concerns about scalability frequently arise with public blockchains because of the volume of transactions and identity information that they process.

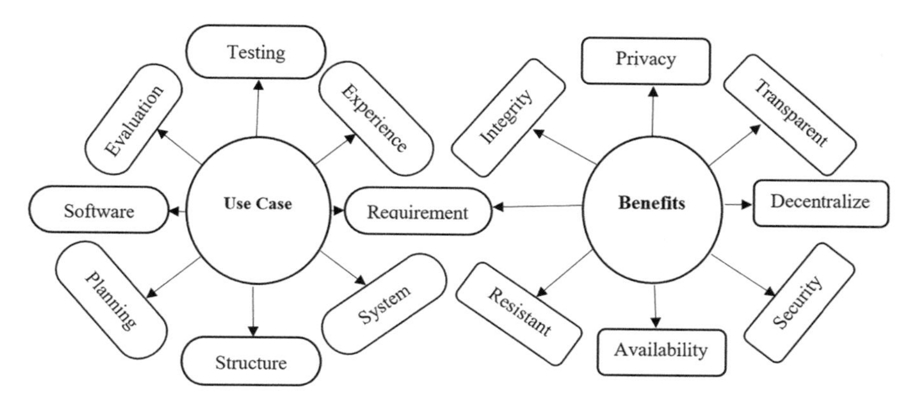

Fig. 8.2 Key features of and use cases for a blockchain

In decentralized identity management, which stores and manages identity attributes, scalability is crucial for efficient and quick verification and authentication. Scalability also stores and manages identity attributes.

Privacy and security: Blockchain technology ensures security through immutability and decentralized consensus; however, maintaining users' anonymity is challenging because of the nature of the platform. A compromise must be struck between protecting one's identity and maintaining transparency. It is challenging to protect the privacy of data and implement privacy-enhancing techniques such as zero-knowledge proofs without compromising the integrity of the system or its credibility [15].

Interoperability: Interoperability is a challenge because it is difficult to integrate blockchain platforms and identity systems. Fluidly sharing data between identity providers, reliant parties, and verification services is an essential component of decentralized identity management. Standards and procedures for interoperability are required in order to transfer identify attributes from one system to another.

Regulatory compliance: There may be problems with regulatory compliance if decentralized identity management is used. The laws governing identification, data protection, and privacy vary widely from country to country and from jurisdiction to jurisdiction. Significant thought and regulatory cooperation are required in order to bring decentralized identity solutions into compliance with these standards while also preserving appropriate governance and liability frameworks [16].

User adoption: It is difficult to change people's mentalities and actions so that they will utilize decentralized identification systems. Users need to be informed about self-sovereign identities, secure digital identity management, and the importance of placing their trust in the underlying technology and procedures. Increasing user adoption requires education, sound guidance, and interfaces that are easy to use.

Infrastructure and integration: Decentralized identity management requires DID registries, verifiable credential issuers, and reliant party systems. This type of management also requires infrastructure and integration. The integration of these components with identification systems and apps is a complex process that needs the coordination of stakeholder interests. Both integration and compatibility are challenging to achieve [17].

Governance and standards: There is no governance model for and there are no standards in place for decentralized identity management. Ongoing challenges include developing governance frameworks that are compatible with the decentralized nature of blockchain identity management and developing agreed-on standards for interoperability, data formats, and verification procedures.

User experiences: User-friendliness drives adoption. Designing intuitive user interfaces and integrating technologies are difficult. Cryptographic key management and decentralized technology might hinder usability. Decentralized identity management promotes privacy but raises concern about data protection and General Data Protection Regulation (GDPR) compliance. Cross-border data flows and the blockchain storage of sensitive personal data make privacy and regulation challenging to reconcile. Integration and acceptance necessitate decentralized identity management solution interoperability. Common standards, protocols, and governance frameworks for identity interoperability across platforms and services are complex and require stakeholder consensus [18].

Traditional centralized identification systems allow identity recovery and revocation. Decentralized identity management, where users have full control over their identities, impedes identity recovery and revocation. Decentralized identity management laws are changing. Digital identity ownership, liability, dispute resolution, and cross-jurisdictional difficulties must be solved to legalize and promote decentralized identity solutions. Trusting decentralized identity management systems is hard. The goal is to encourage individuals, organisations, and service providers to use identity management solutions for the Internet of Things (IoT), as shown in Table 8.1 [19].

Table 8.1 Blockchain for IoT applications and features

Blockchain-based IoT descriptions					
Sr No.	Properties	Technology	Application	Project	Solution
01	Decentralization	Distributed ledger	Smart care	uPort	MyData
02	Immutability	Smart contracts	Smart grid	Idensys	Waypoint
03	Transparency	Cryptocurrency	Smart city	Tradle	Bloom
04	Latency	Consensus Protocols	Smart finance	Idensys	UniqueId

4 Management

4.1 Scalability

Blockchain-based decentralized identity management needs scalability. Scalability difficulties in public blockchain networks may make identity and transaction management challenging. Scaling involves the implementation of decentralized Identity management (DIM), a process aimed at decentralizing identity management. Blockchain transactions are constrained by time, allowing only one transaction per second. Increased identities and transactions may shut down the network and raise transaction fees. Scaling network throughput improves identity management and allows the network to handle more users [20].

Possible answers: State channels or sidechains can offload transaction processing from the main blockchain, increasing network capacity.

Sharding: Shards process independently transactions on the blockchain network. Sharding parallelizes transactions, scaling these networks.

Storage: As identities and credentials grow, so does the need for blockchain storage. This impacts storage costs and availability.

Off-chain storage: Storing identity-related data off chain in decentralized storage networks like the Interplanetary File System (IPFS) reduces blockchain storage while ensuring data availability and integrity.

Data compression and optimization: Compressing and optimizing blockchain data storage reduces storage without compromising anyone's identity. Creating, maintaining, and confirming IDs require significant computational resources. As identities and operations grow, blockchain computations can overwhelm those resources.

- Identity-related techniques and data structures reduce computing overhead.
- TEEs help offload computation, which lessens the blockchain-processing burden.

Interoperability: Decentralized identity management systems must be interoperable, although scaling may be problematic. Identity and credential communication across blockchain networks and services requires scalable and efficient cross-chain communication protocols. DIDs, VCs, and W3C DID and VC standards can help decentralized identity systems integrate and interoperate. Efficient cross-chain communication techniques for secure and scalable identification data transmission can improve interoperability without sacrificing scalability.

4.2 Privacy and Security

Blockchain-based identity management needs privacy and security. Blockchain is secure, but decentralized identity management privacy and security must still be considered.

Selective disclosure promotes privacy: Verifiable credentials allow people to exchange only transactional information and to do so without revealing personal information. Minimal disclosure protects privacy. Blockchain-based decentralized identity management allows anonymity, and even pseudonymity protects private information.

Off-chain storage: IPFS or encrypted storage can protect private data. This safeguards sensitive identifying data.

Zero-knowledge proofs: These proofs can verify a claim without needing data. Cryptographic methods authenticate identification without revealing attribute values.

Safety issues: Decentralized identity management requires strong cryptography. Public-key cryptography authenticates identities. Identification data are encrypted and signed. Key management safeguards decentralized identities. Users must safeguard their identity-controlling private keys. Hardware wallets, multifactor authentication, and secure key storage help to mitigate key risks. Blockchain's immutability safeguards identity-related transactions and data. Changes to blockchain identity information require authority [21].

Consensus mechanisms: The blockchain network's consensus process should be secure to avoid identity data modifications or attacks. Proof of Work (PoW) and Proof of Stake (PoS) consensus algorithms secure blockchain networks. Decentralized identity management requires blockchain network security. Identity data require network-level encryption, safe node connectivity, and Distributed Denial of Service (DDoS) avoidance. Blockchain allows identity management audits. Tracking and validating identification data transactions and modifications improves accountability [22].

4.3 Interoperability

Blockchain-based identity management requires compatibility. Decentralized identity systems, platforms, and services easily share and analyze identity-related data, credentials, and interactions. Interoperability is essential for a connected, efficient ecosystem that trusts identities across domains. Decentralized identity management interoperability requires specific protocols, technologies, and methods.

Standard protocols: Interoperability requires standardization. W3C standards for decentralized identity management include DIDs, VCs, and DID authentication. Standards help systems to read identity-related data uniformly. Cross-chain communication between blockchain networks or decentralized identity platforms promotes interoperability. ILP and Polkadot exchange identity-related data and transactions across chains for smooth interoperability [23].

ID bridge technologies: Identity bridges share information across identity systems. Bridges that transfer credentials and attestations promote network interoperability. Chainlink, Sovrin Bridge, and Aries Interop bridge identities. Systems must exchange identification data and terminology for semantic interoperability. JSON-LD standards, for example, provide semantic data interpretation and allow for the exchange of identity features and credentials. Interoperability requires blockchain developers, identity management experts, standards groups, and regulators. Governance mechanisms and industry consortia regulate interoperability standards. Plug-and-play solutions for decentralized identification systems promote interoperability. These solutions should follow common standards and offer interoperability out of the box, allowing enterprises and service providers to seamlessly join and engage with decentralized identity networks [24].

User-centric methods: User identities should be interoperable across platforms and services. Users should manage, control, and selectively disclose their identity attributes regardless of the decentralized identification system. Interoperability should comply with regulations like GDPR. Interoperable data management technologies are needed to maintain user confidence and to comply with privacy rules.

4.4 User Adoption

Users drive blockchain-based decentralized identity management. Adoption limits the benefits of decentralized identities.

Usability/user experiences: Easy user uptake is needed. Decentralized identity management systems should include simple setup, onboarding, and use instructions. Reducing technical complexity and integrating user activities improve usability.

Education and awareness: Users must grasp the value of decentralized identities. Few people understand self-sovereign identification and decentralized identity management. Real-world use cases and success stories may convince skeptics and encourage adoption. Benefits and incentives may boost the adoption of decentralized identity solutions. These incentives reduce transaction costs and improve privacy, security, service access, and identity verification. Demonstrating how decentralized identity management simplifies and protects users can increase adoption. User adoption necessitates collaboration between service providers and businesses. Decentralized identity systems in online marketplaces, social networks, and banks benefit users. If decentralized identification enhances user experiences and enables new functions, service providers may adopt and promote it [25].

Confidence and security: Decentralized identity management solutions require user confidence. Security measures, user control, data protection, and transparency about technology and protocols increase user trust. Third-party audits, certifications, and privacy compliance ensure system security and compliance. User adoption requires integration with existing identity systems and infrastructure. Single Sign-On (SSO) lets consumers use their digital identities while moving to decentralized identity management. User adoption requires standardization and industry cooperation. When systems and platforms follow standards, users can more seamlessly connect with multiple services and businesses can integrate decentralized identity solutions. Standardization and industry consortia accelerate the development and implementation of decentralized identity management. User trust demands GDPR compliance. Demonstrating that decentralized identity management solutions respect privacy and rules can enhance user confidence [26].

4.5 Regulatory Compliance

Blockchain-based identity management needs regulation. Decentralized identity systems handle sensitive data, so legal compliance is essential. Regulations on decentralized identity management must protect users from several concerns.

Data security: Many nations have data-privacy legislation. The European Union (EU) created the General Data Protection Regulation (GDPR), which governs data processing and protection. Decentralized identity management solutions must comply with these rules and lawfully handle user data.

Minimizing personal data: Collecting and processing must use the least number of personal data needed for identity management and must minimize data retention.

User consent: Users must be informed and express consent before their personal data are collected and processed. Data-processing objectives must be defined.

Data subject rights: These rights guarantee access to, corrections to, and the deletion of personal data. Data subject requests must be optimized. Decentralized identity management systems may require identity verification and KYC. Financial compliance and healthcare compliance are strict to prevent fraud, money laundering, and identity theft.

Risk-based approach: Transaction and user risks must be identified.

Sector-specific rules: Anti-Money Laundering (AML) and Counter-Terrorism Financing (CTF) requirements may demand further identification verification; users must understand this and comply.

Jurisdictional compliance: Decentralized identity management companies must follow local laws and regulations. Country-specific rules govern personal data processing and transfer.

Cross-border data transfers: Use data transfer safeguards like Standard Contractual Clauses (SCCs) or Binding Corporate Rules (BCRs). when moving personal data across borders.

Data localization needs: Data localization requirements may restrict personal data storage and processing outside particular geographical borders.

Crisis: Identity data must be protected, so responses to security breaches and data breaches must be planned.

Monitor regulatory changes and communicate with regulatory agencies or industry bodies to understand decentralized identity management needs and expectations.

Compliance must be maintained, and updates must be processed. Complex decentralized identity management regulatory compliance is evolving. Legal counsel and regulatory bodies should advise organizations employing such technologies to comply with local legislation [27].

5 Solutions to Address Challenges

Blockchain-based identity management is hard, requiring layer 2 scaling.

Layer 2 scaling: State channels, sidechains, and off-chain technologies increase the number of transactions and decrease blockchain network pressure.

- For parallel transaction processing and scalability, shard the blockchain network.
- Optimized consensus mechanisms such as PoS or dPoS improve performance and scalability.
- For privacy/security issues zero-knowledge proofs can be used for private, data-free verification.
- To ensure encryption and confidentiality, only authorized parties can access sensitive identity data stored on the blockchain or in off-chain storage options.

Hardware wallets, multifactor authentication, and safe storage protect private keys and identities.

Auditable smart contracts: Smart contracts should undergo rigorous security evaluations to find and fix problems.

Interoperability challenges: DIDs, verifiable credentials, and W3C-developed decentralized identity-related standards can be used to overcome interoperability challenges. In this way, interoperability and identity data exchange can be improved [28].

Identity bridge technologies: Interoperable decentralized identity systems and blockchain networks use identity bridge technologies to exchange credentials and attestations.

Industry stakeholders, developers, and standards groups should be encouraged to collaborate and deploy suitable solutions.

User-adoption challenges: To improve user experiences, create simple interfaces and onboarding for decentralized identity management solutions. Simplify setup, and give clear instructions.

Education and awareness: Inform users of decentralized identity management's security, privacy, and other benefits. Promote awareness, education, and use cases to highlight decentralized identity's benefits.

Service providers: Service provides should integrate decentralized identification solutions, and the ways that decentralized identity can benefit service providers and users should be demonstrated to them.

To promote decentralized identity systems, users should be offered lower transaction costs, better privacy, or exclusive access to services.

Compliance issues: Consider GDPR and privacy while designing decentralized identity management solutions. Start development with privacy, data reduction, and user approval.

Collaboration with regulatory bodies and authorities: Understand and comply with decentralized identity management laws and regulations. Decentralize identity regulation frameworks.

5.1 Scalability Solutions

Blockchain identity management needs scalability. Scalability is a problem with many potential solutions.

Layer 2 scaling: Layer 2 scaling features off-chain transaction processing with blockchain security. State channels, sidechains, and off-chain protocols work. Off-chain identity-related transactions improve scalability by recording just the final judgment on the main blockchain.

Sharding: Sharding splits the blockchain network. Each shard independently handles transactions, increasing network throughput. Sharding parallelizes transactions, reducing congestion and improving scalability. Sharding decentralized identity systems allows for more identity-related transactions to be carried out. Consensus procedures also scale blockchains. PoS/dPoS consensus mechanisms have higher throughput and scalability, and these consensus methods can scale decentralized identifying systems and require less processing power than PoW does [29].

Off-chain storage: Storing plenty of identity data on the blockchain can pose scalability concerns. Distributed file systems or decentralized storage networks store identity data off chain. These systems store identity data safely and cheaply.

Batch processing: Batch processing combines identity-related blockchain transactions. This reduces on-chain transactions, improving scalability. Batch records contain multiple identity verifications or credential issuances, reducing latency and transaction throughput by optimizing network architecture. Network partitioning, data compression, and efficient peer-to-peer communication protocols improve decentralized identification system efficiency and scalability. Blockchain protocols must be updated. Ethereum 2.0 scales with shard chains and speedier consensus. Decentralized identity management is subject to upgrades [30].

Continuous R&D: Scalability needs constant innovation. Scalable identity management protocols and solutions should be built with researchers from the decentralized identity community.

5.1.1 Off-Chain Storage

Blockchain identity management uses off-chain storage. Storing massive volumes of identity data on the blockchain is wasteful and limits scalability. Storing identity data off chain is cheaper and more scalable. Off-chain storage systems store data and both secure and verify identity data [31].

Decentralized storage: IPFS and Swarm segment data, distribute them between nodes, and ensure data availability.

- Blockchain-based storage marketplaces include Filecoin and Sia. Renting network storage guarantees data durability and redundancy.
- Amazon S3 and Google Cloud Storage store identification data off chain. Blockchains store cloud data references and cryptographic evidence.
- Off-chain storage helps blockchains scale by reducing the number of on-chain data. Decentralized identification systems can manage more transactions and users.

Cost-effectiveness: Computational and storage requirements make blockchain data storage expensive. Off-chain identity data storage is cheaper.

Flexibility: Off-chain storage solutions offer a variety of data formats, protocols, and access controls. They can save images, videos, and papers for identity data.

Privacy: Off-chain storage options keep personal data off the blockchain. Off-chain data are secure and confidential, while on-chain data are just cryptographic proofs.

Blockchain security: Blockchains secure off-chain data. Merkle trees, digital signatures, and hash pointers enable this link. These methods audit off-chain data by using blockchain data.

Tradeoffs and considerations: Off-chain storage offers affordability, scalability, and tradeoffs. Storage providers safeguard off-chain data. Users should choose only reliable storage networks.

Data availability: Off-chain storage systems need redundancy and availability to prevent data loss and must encrypt data and restrict off-chain data access to only authorized parties.

Synchronization and consistency: Keep blockchain and off-chain identifying data consistent. Decentralized identity management can handle huge identity data with off-chain storage. Off-chain storage saves money, scales decentralized identity systems, and protects privacy [32].

5.1.2 Sharding and Layer 2 Solutions

Sharding and layer 2 enable blockchain-based decentralized identity management.

Sharding: Sharding splits the blockchain network. Each shard performs transactions independently and simultaneously.

Benefits of sharding: Sharding lets multiple shards process transactions simultaneously, increasing network throughput. Decentralized identity systems handle a greater volume of identity-related transactions compared to centralized alternatives, and sharding optimizes efficiency by distributing transaction loads across multiple shards, reducing network congestion and improving efficiency. This scales with the speed of transaction confirmations.

Shards process transactions individually. Decentralized identification systems scale horizontally.

Layer 2: Layer 2 refers to blockchain-secured off-chain transaction processing. On-chain transactions can be reduced to scale.

Layer 2 decentralized identity management methods: State channels allow off-chain transactions without blockchain recording. Blockchain transactions are private and safe. Identification verification uses state channels. Sidechain identity transactions lighten blockchains, and sidechains are faster and more customizable for decentralized identity systems.

Off-chain protocols: Bitcoin's lightning network offers secure, scalable off-chain transactions. Participants use payment channels to settle off-chain transactions on the main blockchain. Off-chain identity interactions scale. Layer 2 solutions scale decentralized identification systems by off-chaining transactions. These solutions improve identity-related transaction confirmation speeds, reduce costs, and increase scalability. However, sharding and layer 2 have perks and cons. Decentralized identity management systems must be properly planned, implemented, and tested to ensure data confidentiality, integrity, and consistency and to improve scalability [33].

5.2 Privacy Solutions

Blockchain identity management demands secrecy.

Privacy options: Off-chain storage solutions keep sensitive identifiable data off the blockchain. Off-chain encryption provides privacy. On-chain data access and sharing improve privacy.

ZKPs: Zero-knowledge proofs use cryptography to verify a statement without revealing any facts. ZKPs can validate identities without revealing personal information, thanks to decentralized identity management. They enable users show their qualifications without sharing their sensitive data. Differential privacy makes query responses and statistical analysis noisy, making data points hard to identify. Aggregating and analyzing identification data in decentralized identity systems protects user privacy. ZKPs encrypt blockchain or off-chain identification data, and encryption protects data. Symmetric, asymmetric, or homomorphic encryption protects identity data.

Data minimization: Only important identity-related data should be stored *on* the blockchain. Data minimization reduces privacy breaches. Decentralized identity systems should store only those data needed for specific interactions and should keep sensitive data off chain. Privacy-protecting smart contracts can privately carry out identity-related computations by using state channels or encrypted data structures. Users should control their identity data and carefully choose which credentials to share, with whom, and why. Transparent consent and user-friendly interfaces enable privacy control [34].

Privacy by design: Decentralized ID systems start with privacy. To prioritize privacy throughout the system's lifecycle, privacy-enhancing methods and best practices should be followed and privacy impact evaluations should be carried out.

5.2.1 Zero-Knowledge Proofs

Zero-knowledge proofs (ZKPs) use cryptography to prove a proposition to a verifier without giving any additional information. ZKPs increase blockchain-based identity management privacy and secrecy and have several other uses.

Identity authentication: ZKPs can authenticate an identity without revealing it. Users can prove they're at least 18 years old without disclosing their respective birthdays. ZKPs enable the prover to prove characteristics or circumstances without revealing sensitive information. ZKPs allow selective identity-related disclosure. Users can verify themselves anonymously and can confirm their academic degree without revealing their university or course. ZKPs authenticate knowledge without revealing it. A verifier can verify users without revealing network credentials. This reduces credential theft and interception [35].

Privacy-preserving transactions: ZKPs can be used to prove whether participants have sufficient cash or meet certain conditions, without disclosing their transaction history or the amount transferred. This safeguards financial transactions. ZKPs preserve input privacy for safe multiparty computing. This may help with complex identity-related data operations like aggregating statistics or data analysis in decentralized identity management without revealing individual data points.

Credential revocation: ZKPs can authenticate credential revocation without revealing the credentials or user privacy. This detects revoked credentials and blocks their use.

5.2.2 Confidential Transactions

Blockchain-based decentralized identity management uses cryptographic confidential transactions to protect privacy. Confidential transactions hide blockchain transaction quantities. Traditional blockchains show transaction amounts to all participants. Confidential transactions encrypt amounts for only the intended recipients while allowing the network to validate the transaction.

Pedersen commitments: Cryptographic frameworks hide and verify value in confidential transactions. Pedersen commitments blindly encrypt transaction amounts. The blockchain verifies this promise without revealing the amount.

Range proofs validate confidential transactions. Cryptographic range proofs confirm that the committed value is positive or within a given maximum value. Range proofs enable the network to verify transaction amounts without releasing them. Decentralized identity management hides identity-related transaction quantities with confidential transactions. To prevent identity-related behavior-inference or behavior-correlation attacks, identity verification systems can keep transaction amounts private. Confidentiality protects decentralized identity management systems during financial transactions. Participants can trade without disclosing account balances or transfer amounts. This safeguards identity management ecosystems during financial transactions [21].

Privacy-safe smart contracts: Confidential smart contracts protect sensitive data. Smart contracts may compute secret transaction amounts without disclosing them, maintaining blockchain integrity and privacy in complex procedures. Decentralized identity management protects privacy with confidential transactions. Confidential transactions let users privately carry out identity-related tasks while they benefit from blockchain transparency, immutability, and security.

5.3 Interoperability Solutions

Interoperability lets blockchain-based decentralized identity management systems and networks share identify data.

Interoperability Standards

Interoperability standards allow decentralized identification systems to share data. DIF, W3C verifiable credentials, and DID standards promote identity management solution interoperability.

Universal identifiers: Decentralized identifiers (DIDs) standardize blockchain identity descriptions. DIDs enable blockchain and decentralized identity platform interoperability by recognizing and referencing identities. Cross-chain interoperability lets blockchain networks share assets, including identity data. Atomic swaps, sidechains, and interoperability protocols (e.g., Polkadot and Cosmos) allow data and assets to smoothly migrate across blockchains, boosting decentralized identification system compatibility. Blockchain networks connect by using interchain communication protocols. Protocol-level interoperability is enabled by secure, trustless interblockchain communication (IBC). Interblockchain communication protocols transmit identification data and credentials across blockchain networks [36].

Bridge solutions: Bridge solutions allow the interoperation of dispersed identity systems. These bridges allow systems to communicate and exchange identity-related data between formats, protocols, and standards. Developers can combine many decentralized identification systems by using open Application Programming Interfaces (APIs) and Software Development Kits (SDKs). Open

APIs help identity management platforms to share data. Blockchain platforms, identity suppliers, and standardization organizations must collaborate for interoperability. Collaboration may involve standardization, best practices, and cross-platform compatibility. Decentralized identification systems can overcome siloed approaches by using interoperability solutions. Interoperability makes decentralized identity management more useful, scalable, and successful by integrating the digital identity ecosystem [37].

5.3.1 Cross-Chain Communication

Blockchain-based identity management requires cross-chain communication for interoperability. Blockchain networks can share identity data and transactions. Atomic swaps enable blockchains to trade assets, including identity data, without a trusted intermediary. Users can securely transport currencies or data across blockchains without carrying out centralized exchanges.

Pegged assets: Sidechains are interoperable. Transferring identity-related data and assets between the main blockchain and sidechains enables scalability and specialized services. Backing tokens on one blockchain with assets on another confers cross-chain value and allows data exchanges. interoperability protocols such as Polkadot, Cosmos, and Aion connect blockchains. These protocols allow blockchain transactions, data sharing, and decentralized identification system interoperability. Interblockchain communication (IBC) lets Cosmos blockchains interchange data. It secures and decentralizes identification data interchange between blockchains.

Wrapped tokens: One blockchain represents another. They confer cross-blockchain value and enable data transactions. Wrapped tokens link blockchain credentials and reputation scores. Oracles link blockchains to external data. Smart contracts and decentralized apps leverage their data. Oracles can connect blockchain networks and add identification data. Cross-chain smart contracts execute logic and actions. They allow for identifying transactions between blockchain-based decentralized identity systems. Cross-chain communication solutions move identity-related data, assets, and transactions between blockchain networks for decentralized identity management. Security, consensus, and governance are needed to protect identity-related interactions across blockchains [29].

5.3.2 Standardization Initiatives

Standardization aids blockchain-based decentralized identity management and interoperability. Field standardization requires the DIF to create interoperable decentralized identification standards and protocols. The DIF specifies DID, verifiable credentials, and DID authorization. Standards enable decentralized identity management and system interoperability. W3C standards are global, and W3C standardizes decentralized identity technology. This group developed the verifiable credentials specification for issuing, verifying, and sharing digitally signed credentials. Linux Foundation open-source Hyperledger Indy has also created a decentralized identity platform. It provides decentralized identity management tools, libraries, and protocols. Hyperledger Indy's Aries framework supports identity system compatibility and safe peer-to-peer connectivity.

Internet identity workshop (IIW): The community-driven IIW gathers people and organizations working on decentralized identity and related technologies. IIW standardization can be explored by industries, researchers, and practitioners.

European Blockchain Services Platform (EBSI): This EU initiative offers a reliable and interoperable blockchain platform for public services. EBSI's eSSIF covers identity management standards and specifications.

The nonprofit InterWork Alliance (IWA) has created tokenization, smart contract, and blockchain interoperability standards. The IWA may affect multichain identity management. Developers, researchers, and stakeholders establish specifications, protocols, and best practices. Standards enable decentralized identity management interoperability, adoption, and consistency. These projects work together to develop blockchain-based decentralized identities [38].

5.4 User-Adoption Solutions

Users drive blockchain-based decentralized identity management through their acceptance, which is facilitated by several features.

User-friendly interfaces: Intuitive interfaces retain users. Decentralized identification apps should require no technological expertise. User education is needed, however, for decentralized identity acceptance. Consumers may learn to understand decentralized identity management through workshops, webinars, and education. Incentives can encourage decentralized identity solution adoption and use. Token-based loyalty schemes boost ecosystem participation.

Collaboration: Decentralized identification solutions can help people embrace existing platforms and processes. Integration with popular apps, social media, and financial institutions can attract users and simplify decentralized identity.

Privacy and data ownership: Privacy-conscious consumers would like this. Decentralized identity's data control and privacy can boost adoption.

Pilots and use cases: Successful pilots and real-world use cases demonstrate decentralized identification. Effective decentralized identification solutions for healthcare, supply chains, and finance can enhance confidence and acceptance.

Industry partners: Governments, businesses, and service providers increase user adoption. Strategic partnerships can increase the value of decentralized identity as more organizations and people join the ecosystem.

User assistance and feedback: Reliable user help and active user feedback improve the user experience and fix issues. User feedback promotes advancements to decentralized identity solutions. Regulations and standards increase user trust. Data security, data privacy, and industry standards improve the legitimacy and adoption of decentralized identification systems.

Scalability and performance: Large user and transaction counts necessitate scalable and performant decentralized identity solutions. Users choose technologies that meet their needs quickly and smoothly.

5.4.1 User-Centric Interfaces

User-centric interfaces boost blockchain-based decentralized identity management adoption and experience. User-centered interface design has several important features.

Easy onboarding: Decentralized identity setup should be easy upon onboarding. Clear explanations and minimal technical jargon facilitate basic setup. Consumers' mental models must align with sensible user flows. The decentralized identity management system should have an intuitive User Interface (UI) and should avoid information overload. Visual cues, icons, and tooltips provide context and interface direction. Users' credentials and transaction information should easily be displayed.

Visualizations: Complex decentralized identity ideas should be explained with diagrams or flowcharts. Visuals can explain blockchain identification and data management. Visuals must allow interface customization for layouts, colors, and notifications. Ownership and customization let users access the decentralized identity management system [14].

Transparent control: Users need to be able to view and manage their identity-related data, so transparent controls must allow users to choose which identity information to share and with whom. These controls prioritize consent and make access revocation easy.

Mobile-friendly design: Such a design creates responsive interfaces on various devices for the users of decentralized identity management systems. For consistency, mobile apps and responsive web interfaces should be tailored to being used on smaller screens.

Help: User-centered interface design offers significant interface-based supports, such as frequently asked questions (FAQs), knowledge bases, chatbots, and customer service. To improve user confidence in the decentralized identity system, it must swiftly respond to user inquiries and provide clear support.

Usability testing and iterative design: Such testing provides user feedback, which will inform future improvements. An iterate UI is based on user feedback. To generate a user-centered design, always test and validate it.

Accessibility: A multiuser interface improves user accessibility. Consider color contrast, typeface size, screen reader compatibility, and keyboard navigation. Accessible interfaces promote adoption.

5.4.2 Education and Awareness

Blockchain-based identity management requires education and awareness. There are several awareness-raising methods available.

Educational resources: Decentralize identity management concepts, benefits, and use cases via whitepapers, manuals, tutorials, and movies. Technical and non-technical audiences should find these items useful.

Workshops and training: Decentralized identity management workshops, seminars, and training are available for individuals, corporations, and organizations. These events offer real use cases, demonstrations, and hands-on experiences.

Online forums: Decentralized identity management aficionados should have online groups and forums for them to discuss, learn, and share ideas. Participation, cooperation, and best-practice sharing boost learning on these platforms.

Industry conferences: Identity, blockchain, and digital identity conferences should present, conduct panels, and engage participants to promote decentralized identity management and its benefits. Professionals network at these events. Decentralized identity management should be integrated into academic courses and university-research cooperation. This connection can nurture future specialists and innovators and can improve academic–industry information transfer. Success stories on decentralized identity management should be showcased. These examples demonstrate the technology's applications, benefits, and consequences, encouraging others to explore it [39].

Tech communities: Developer networks and technology forums need to reach techies. Developer guides, code samples, and SDKs incorporate decentralized identity management into new and existing applications. Developers can improve technology by experimenting. Social media, blogs, podcasts, and online publications can be used to promote decentralized identity management; useful articles, interviews, and success stories promote the technology; and working with digital identification and blockchain technology industry alliances, standards bodies, and advocacy groups also promotes the technology. Industry-wide initiatives, working groups, and standardization increase awareness, align best practices, and establish a cohesive ecosystem [40].

Public education: Educating the public should promote decentralized identity management. These advertisements support self-sovereign identification, data privacy, and security and warn against centralized identity systems.

5.5 Regulatory Compliance Solutions

Blockchain identity management involves regulatory compliance. Although regulatory compliance comes with problems, several solutions are available.

Comply with laws: GDPR, identity verification, and sector-specific compliance standards should be monitored. Decentralized identity management solution must meet these laws' exact requirements.

Privacy by design: Privacy must be built into any decentralized identity management system. Privacy regulations require data reduction, consent management, and purpose limitation. They must encrypt and pseudonymize user data.

Data governance and consent management: User data must be handled with effective data governance, such as by following data ownership, storage, and access policies. User consent management technologies let people control their data and make informed decisions about their use and dissemination.

Compliance auditing and reporting: Any decentralized identity management system's regulatory compliance must be regularly examined. Records must be maintained to establish compliance and report to regulators.

Secure and immutable audit trails: Blockchain technology's transparency and immutability enable secure, tamper-proof audit trails. These audit trails track user interactions, credential issuances, and consent management to ensure compliance. Any decentralized identity management system should have strong identity verification and anti–money laundering protocols. Fraud prevention requires AML-compliant user onboarding, identity verification, and transaction monitoring. Compliance requirements can be discussed with regulatory agencies and authorities. Joining regulatory compliance industry alliances and working groups can help [41].

Compliant smart contracts: Decentralized identity management smart contracts must meet legal and regulatory requirements. Contracts should stipulate compliance.

Legal issues: Knowing how a decentralized identity management system affects jurisdiction will help when adjusting to them. Identity, data, and privacy regulations vary by country and jurisdiction, so every system must be adjusted accordingly.

Monitoring and adjusting: Compliance legislation changes. Every decentralized identity management system should meet the new regulations. When a compliance program's rules change, adjust the system to maintain compliance.

5.5.1 Collaboration with Regulatory Bodies

Blockchain-based decentralized identity management needs regulatory cooperation, which necessitates communicating with decentralized identity management regulators. Government, regulatory, and industry-specific identity management and data-privacy organizations must be considered when working with regulators.

Join regulatory consultations: Participate in digital identification, blockchain, and data-protection regulatory working groups. Consult industry experts on decentralized identity management regulatory frameworks. Discuss regulations and blockchain-based identity systems [5].

Request regulations: Ask regulators about decentralized identity management compliance and best practices. Discuss regulatory requirements and the technical capacity to comply.

Educate regulators: Inform authorities on blockchain-based decentralized identity management's benefits, technicalities, and prospects. Explain how the technology improves privacy, security, and user control. Help regulatory agencies to understand and address decentralized identity management.

Pilot programs: Pilot or test decentralized identity management technologies in regulatory sandboxes. These applications simulate blockchain-based identity system regulations and practicalities. They allow regulator–innovator collaboration and real-world testing.

Proportional regulation: Balance innovation with consumer protection regulations. Decentralized identity management can address privacy, security, and user empowerment, but flexible regulatory procedures are needed to enable innovation, interoperability, and market competitiveness.

Regulatory compliance: Create a decentralized identity management system that meets regulations. Comply with regulators and apply system controls. Compliance conversations help to ensure regulatory compliance.

Transparency and regulatory reporting: Alert regulators. Generate reports, respond to regulatory enquiries, and resolve compliance issues swiftly. Decentralized identification systems are trustworthy when transparent and regulated.

Industry standards and certification: Create decentralized identity management standards and certification frameworks with regulatory authorities and industry associations. Standards and certification collaboration ensure system and provider standardization, interoperability, and compliance. Actively monitor compliance in any decentralized identity management system. Internal audits, proper compliance records, and regulatory compliance analysis are essential. Fix compliance issues [42].

5.5.2 Adaptable Frameworks

Blockchain-based decentralized identity management must adapt to user, enterprise, and regulatory needs. Making adaptable frameworks requires attending to several considerations.

Modular architecture build: Use a modular architecture build to produce a modular, decentralized identity management platform. Modularity lets the system adapt to changing needs without disrupting any infrastructure.

Standards adherence: Integrate with various identity management systems by following industry standards and compatible protocols. Standardization promotes interoperability, cross-platform communication, and decentralized identity.

Plug-and-play parts: Create a plug-and-play framework for users and organizations to choose and integrate components. Customers can modify and scale, adding features as needed. Customize the system with framework configuration options. Privacy, consent, authentication, and identity verification are included. Established user settings ensure privacy and regulatory compliance.

Governance and consensus systems: Establish governance mechanisms for decision-making and community involvement. Stakeholders can influence identity management through consensus, voting, or Decentralized Autonomous Organizations (DAOs).

Upgradability and compatibility: Support new blockchain, identity, and cryptography standards. The framework can adapt to technical, security, and regulatory changes without rebuilding.

Legality: Regulate framework design and architecture. Data protection, identity verification, and compliance are included. The framework's flexibility permits regulatory-specific compliance.

Cooperative feedback: Facilitate framework input to engage users, developers, and stakeholders. Examples of feedback input includes user input, developer communities, and open-source contributions. Community input improves the framework.

Developer tools: Provide detailed documentation, developer resources, and APIs to integrate and develop applications and services on the framework. Clear documentation lets developers personalize the framework.

Trials and pilots: Test and pilot the framework's functionality, usability, and flexibility in the real word. User, organization, and regulatory agency comments can improve the framework [43].

6 Case Studies and Existing Implementations

Decentralized identity management case studies use blockchain.

Sovrin: Blockchain-based Sovrin gives global self-sovereign identity. Individuals and organizations can govern their digital identities with its decentralized identification infrastructure. Sovrin protects identity interactions with DIDs, VCs, and ZKPs.

uPort: This Ethereum-based self-sovereign identity manages digital identities and personal data and selectively shares information with service providers. Decentralized Identifiers (DIDs), Distributed Key Management System (DKMS), and smart contracts personalize identity management for uPort.

Microsoft's DID framework: W3C standards underpin Microsoft's DID framework. Users can create, own, and control their digital identities for privacy and interoperability. Microsoft supports blockchain, identity hubs, and decentralized key management.

Civic: Civic authenticates identities and shares personal data with trusted entities by using blockchain technology. Civic's decentralized identification enables users to save and share their identity information on their devices. Civic uses blockchain to safely and openly verify IDs.

Verity: Evernym's decentralized identity platform allows organizations to issue and verify VCs. Sovrin's distributed ledger secures and interoperates self-sovereign identification for Verity. It manages passwords, data, and privacy-enhanced disclosures [44].

6.1 Identity Management in Healthcare

Blockchain-based decentralized identity management benefits healthcare. Healthcare blockchain-based decentralized identification management has several key features.

Blockchain protection: Blockchain protects patient identity verification. Decentralized ledgers protect patient data against identity theft and fraud. Patients can easily share verified information with healthcare institutions and control their identities. Blockchain-based decentralized identity management interoperates medical records. Patients may access and share their medical records among providers with a single digital identity, eliminating data entry and improving care coordination. While sharing, blockchain secures patient data. Blockchain technology enables patients to give fine-grained data-sharing consent. The blockchain allows patients to choose which healthcare institutions can access their medical data and for what purposes. This enhances patients' data-sharing decisions and safeguards their privacy [45].

Clinical trials and research: Blockchain-based decentralized identity management simplifies patient recruitment. Securely combining patients' identities with their health data on the blockchain lets researchers quickly identify qualified participants while maintaining patients' privacy and data integrity. Patients may thus entrust their data to research.

Prescription and medication management: Blockchain enhances security and traceability. The blockchain can link patients' digital identities to their pharmaceutical records, reducing medication errors, counterfeit drugs, and supply-chain tracking. Blockchain-based identity management protects healthcare data. Only authorized parties can decrypt blockchain-stored patient IDs and sensitive health data. Decentralized identities (DIDs) and verifiable credentials (VCs) increase privacy by reducing the level of personal data disclosure.

Blockchain-based identity management detects and prevents healthcare fraud. A secure patient identity and healthcare transaction record helps detect fraud. Blockchain's transparency allows auditors and authorities to investigate suspicious activities and protect the healthcare system. Decentralized identity management allows patients to have more control over their health data. Patients can choose healthcare providers, revoke access to their data, and view data use. Patient trust and healthcare engagement therefore increase. Decentralized identity management must involve healthcare providers, technology vendors, regulators, and patients. Legal, regulatory, data-protection, and healthcare industry norms and sensitivities must be addressed. Blockchain-based decentralized identity management could improve patient care, data security, and interoperability in the healthcare industry [46].

6.2 Digital Identity for Financial Inclusion

Financial inclusion is possible with blockchain-secured digital IDs. Decentralized identity management helps financial inclusion in several ways:

Identity verification: Blockchain-based decentralized identity management lets undocumented people construct digital identities. Financial businesses can verify IDs by using blockchain's immutability and cryptography. This helps people who lack official documentation or who have limited identity verification systems.

Financial services: Decentralized identity management enables remote access to financial services. Blockchain-based identities establish trust and trustworthiness, enabling bank accounts, loans, and formal financial participation.

Cross-border payments: Blockchain-based identities enable faster, cheaper cross-border payments, especially for those without bank accounts. Decentralized identities offer fast, secure identity verification and cross-border transactions. Decentralized identity management improves microfinance and peer-to-peer lending networks. Blockchain-based identities help financially excluded people to access microloans and peer-to-peer loans and showcase their creditworthiness [47].

Transfers: Blockchain-based digital IDs simplify and safeguard underserved transfers. By establishing their identities on the blockchain, individuals can verify their eligibility to receive remittances, reducing transaction costs and expediting and securing cross-border transactions.

Financial data privacy and security: Blockchain-based decentralized identity management gives people control over their financial data. Self-sovereign identities allow individuals to choose to share financial information to financial service providers, retaining data privacy while accessing crucial services. This safeguards data and gives users more control. Blockchain-based decentralized identity management simplifies KYC for financial institutions. The blockchain can verify identity once and share it with several organizations, saving time. Financial institutions can meet regulations without burdening consumers. Financial inclusion requires financial institutions, technology vendors, regulators, and local communities to adopt decentralized identity management. Legal, regulatory, data privacy, and underrepresented demographic demands must be addressed. Blockchain technology for decentralized identity management can promote financial inclusion by allowing safe and portable identities to access important financial services and participate in the global economy [48].

6.3 Decentralized Identity for IoT Devices

Blockchain-based decentralized identity management improves IoT-device security, interoperability, and data privacy. Decentralized IoT identity builds trust in using IoT devices:

Device trust: Blockchain-based decentralized identity management secures IoT-device authentication and confers trust. A blockchain-stored DID can identify each device. Authenticating and securely connecting devices, networks, and apps reduces the risk of unauthorized access and device spoofing. Decentralized identities allow IoT devices to safely share data and communicate with other platforms. Verifiable credentials and cryptography protect shared data. Decentralized identity management makes IoT devices and platforms compatible. Identity protocols and blockchain-based IDs allow IoT devices to share data across networks and ecosystems. This simplifies IoT system management. IoT-device owners can control their data by using decentralized identity management. Companies using device data can receive granular approval from owners. Finally, privacy and IoT-device data management are protected [49].

Blockchain-based decentralized identity management can provide supply-chain IoT devices with unique identities, boosting transparency. These IDs can verify the origin, legitimacy, and supply-chain movement of products, boosting stakeholder trust [50]. Decentralized identity management secures IoT firmware updates. Blockchain device IDs allow approved firmware changes. Blockchain immutability allows firmware auditing and verification, reducing malicious manipulation. Decentralized identity management helps IoT systems save electricity. Blockchain identities and cryptographic keys enable device energy optimization and safe access control. Optimization saves energy and money. Blockchain, identity protocols, and IoT platforms enable decentralized IoT identity management, addressing scalability, interoperability, and device resource restrictions. Decentralized identity unlocks the full potential of the Internet of Things by ensuring data integrity, user control, and security for IoT devices [51].

6.4 Cross-Border Identity Verification

Cross-border ID verification is difficult. Blockchain-based decentralized identity management improves the efficiency, security, and privacy of cross-border identity verification. Decentralized identity management grants autonomy. Cross-border verification lets people pick which identification information to share, which enhances data security. Decentralized identification aids cross-border identity verification.

Valid credentials: Blockchain-based decentralized identity management leverages the digitally signed statements of trusted institutions. Names, birthdates, and addresses are verifiable. These credentials can be transferred internationally and confirmed by trusting parties without relying on centralized identity providers, reducing dependence on identification certificates. Blockchain verifies identifying records. Blockchain-stored identity data can lessen the risk of fraudulent IDs or tampering during cross-border verification. Blockchain-based decentralized identity management improves international interoperability and homogeneity. By using similar identity protocols and standards, cross-border verification systems may verify identities across blockchain platforms and jurisdictions [52].

Blockchain-based decentralized identity management lets trusted entities attest to identities. These attestations authenticate identity. Cross-border verification can create confidence and confirm identities through these attestations. Blockchain consensus checks identity. Consensus techniques in cross-border verification systems verify identities and thwart fraudulent information. Zero-knowledge proofs or selective disclosure can protect privacy during cross-border identity verification with blockchain-based decentralized identity management. These technologies allow consumers to verify their identities without disclosing personal information, increasing privacy and lowering data exposure. Collaboration among governments, identity issuers, dependent parties, and technology suppliers is needed for decentralized cross-border identity verification. Legal, regulatory, data-protection, and privacy issues must be addressed to ensure compliance and build trust. Blockchain-based decentralized identity verification could simplify and secure cross-border identity verification, boosting privacy and user control, as shown in Fig. 8.3 [53].

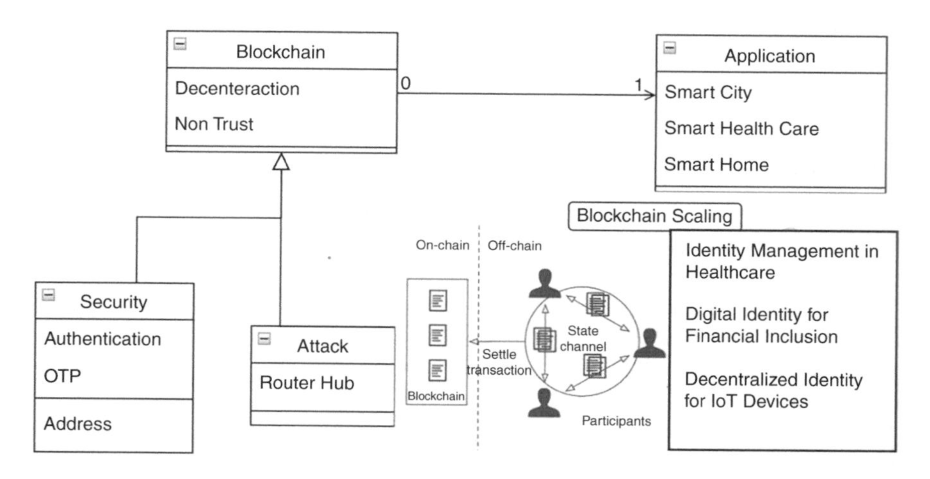

Fig. 8.3 Blockchain transformations in cryptocurrency and the traditional financial world

7 Conclusion and Future Directions

Blockchain-based decentralized identity management can solve identification system issues and offer new avenues for individuals, corporations, and industries. Blockchain's immutability, security, and decentralization enhance identity management's privacy, control, and interoperability. DIDs, self-sovereign identities, and verifiable credentials provide users with more identity control. They can selectively share verified credentials with trusted partners, eliminating centralized identity providers and reducing data breaches and instances of identity theft. Identity management empowers users. Blockchain-based identification solutions can improve security, streamline processes, and enable new business models in healthcare, finance, supply chains, and the IoT. Decentralized identification may affect patient identification, financial inclusion, supply-chain transparency, and IoT-device management.

Making decentralized identity management more popular would require overcoming many difficulties. Scalability, privacy, interoperability, regulatory compliance, and user acceptance matter. Off-chain storage, sharding, zero-knowledge proofs, and regulatory involvement can address these challenges and expand adoption.

Blockchain-based identity management needs further research, standardization, and industry involvement. Consensus algorithms, privacy-preserving approaches, and cross-chain communication protocols improve the scalability, security, and interoperability of decentralized identification solutions. User adoption and compliance require user-friendly interfaces, education, and regulations. Blockchain technology has the potential to transform digital identity management. Decentralized blockchain identity management might empower individuals, improve privacy and security, and enable trusted cross-border transactions.

7.1 Summary of Findings

Blockchain-based decentralized identity management alters digital identities. Identity management is also decentralized. The conceptual framework of blockchain is undergirded by self-sovereign identities, where users can choose to share verified credentials with trusted parties. Decentralized identity management enhances privacy, security, and control. It secures identity verification, promotes platform and jurisdiction compatibility, and removes centralized identity suppliers. Healthcare, finance, supply-chain, IoT, and cross-border identity verification use can be leveraged for decentralized identity management. Decentralized identity management promotes financial inclusivity, supply-chain transparency, secure service access, and IoT-device connectivity. Decentralized identity management challenges include scalability, privacy, interoperability, regulatory compliance, and user uptake. Off-chain storage, sharding, zero-knowledge proofs, standardization, user-centric interfaces, education, and regulatory collaboration are needed to overcome these challenges.

Decentralized identity management has numerous solutions, namely sharding, layer 2 scaling, and off-chain storage scaling. Confidential transactions and zero-knowledge proofs promote privacy and security. Standards and cross-chain communication enable interoperability; user-centric interfaces and education encourage uptake; and regulatory body participation ensures compliance. Decentralized identity management needs research, standards, and collaboration. Consensus techniques, privacy-preserving procedures, and cross-chain communication protocols can increase the scalability and security of decentralized identification solutions. Adoption and compliance require user-friendly interfaces, education, and regulatory frameworks.

7.2 Future Research and Development Opportunities

Blockchain-based decentralized identity management research could focus on user experiences, regulatory frameworks, case studies, scales, and decentralized identification solutions:

Studying scales and decentralized identification solutions: Researchers could study solutions for sharding, off-chain storage, and layer 2 that accommodate additional users and transactions without compromising security or performance. Research could improve decentralized identity management privacy and security. Zero-knowledge proofs, homomorphic encryption, and secure multiparty computing protect user data and transactions while speeding up verification and authentication. Decentralized identification systems and platforms need interoperability standards. To facilitate data exchange among blockchain-based identity solutions, future research could create and promote common identity protocols, data formats, and interoperability standards.

User experiences: For widespread use, decentralized identity systems must improve the user experience. Thus, researchers could study user-centric interfaces, intuitive mobile apps, and user-friendly decentralized identity management solutions. Awareness campaigns could boost decentralized identity management acceptance.

Governance and regulatory frameworks: Research could provide decentralized governance and regulatory frameworks for blockchain identity management. Legal issues, liability frameworks, decentralized dispute resolutions, and identity management systems should be included. Integrating decentralized identity management with upcoming technologies could offer new avenues. Research could connect decentralized identities with IoT devices, artificial intelligence (AI), Machine learning (ML), and edge computing to enable secure, privacy-preserving interactions in complex and dynamic scenarios.

Case studies: Industry-wide case studies could improve decentralized identification solutions. Application cases could be explored in healthcare, finance, supply chains, other industries, and government services to identify and learn more about decentralized identity management's issues, benefits, and effects.

References

1. Venkatraman, S., & Parvin, S. (2022). Developing an IoT identity management system using blockchain. *System, 10*(2). https://doi.org/10.3390/systems10020039
2. Prasad, S. N., & Rekha, C. (2023). Block chain based IAS protocol to enhance security and privacy in cloud computing. *Measurement: Sensors, 28*, 100813. https://doi.org/10.1016/j.measen.2023.100813
3. Chandan, A., John, M., & Potdar, V. (2023). Achieving UN SDGs in food supply chain using blockchain technology. *Sustain., 15*(3), 1–21. https://doi.org/10.3390/su15032109
4. Kişi, N. (2022). Exploratory research on the use of blockchain technology in recruitment. *Sustainability, 14*(16). https://doi.org/10.3390/su141610098
5. Kairaldeen, A. R., Abdullah, N. F., Abu-Samah, A., & Nordin, R. (2023). Peer-to-peer user identity verification time optimization in iot blockchain network. *Sensors, 23*(4). https://doi.org/10.3390/s23042106
6. Juneja, G., & Naswa, R. (2023). Bittrack-A decentralized trust based identity and access management approach. *International Journal of Intelligent Systems And Applications In Engineering, 2023*(5s), 368–388. [Online]. Available: www.ijisae.org
7. Bai, P., Kumar, S., Aggarwal, G., Mahmud, M., Kaiwartya, O., & Lloret, J. (2022). Self-sovereignty identity management model for smart healthcare system. *Sensors, 22*(13), 1–25. https://doi.org/10.3390/s22134714
8. Akbar, M. A., Leiva, V., Rafi, S., Qadri, S. F., Mahmood, S., & Alsanad, A. (2022). Towards roadmap to implement blockchain in healthcare systems based on a maturity model. *Journal of Software: Evolution and Process, 34*(12), 1–15. https://doi.org/10.1002/smr.2500
9. Gilani, K., Ghaffari, F., Bertin, E., & Crespi, N. (2022). Self-sovereign identity management framework using smart contracts, *Proc. IEEE/IFIP Netw. Oper. Manag. Symp. 2022 Netw. Serv. Manag. Era Cloudification, Softwarization Artif. Intell. NOMS 2022*. https://doi.org/10.1109/NOMS54207.2022.9789831.
10. Shobanadevi, A., Tharewal, S., Soni, M., Kumar, D. D., Khan, I. R., & Kumar, P. (2022). Novel identity management system using smart blockchain technology. *International Journal of Systems Assurance Engineering and Management, 13*(s1), 496–505. https://doi.org/10.1007/s13198-021-01494-0
11. Wu, A., Guo, Y., & Guo, Y. (2023). A decentralized lightweight blockchain-based authentication mechanism for Internet of Vehicles. *Peer-to-Peer Networking and Applications, 16*, 1340. https://doi.org/10.1007/s12083-022-01442-0
12. Geetha, R., Padmavathy, T., & Umarani Srikanth, G. (2022). A scalable block chain framework for user identity management in a decentralized network. *Wireless Personal Communications, 123*(4), 3719–3736. https://doi.org/10.1007/s11277-021-09310-5
13. Deng, W., Huang, T., & Wang, H. (2023). A review of the key technology in a blockchain building decentralized trust platform. *Mathematics, 11*(1). https://doi.org/10.3390/math11010101
14. Taherdoost, H. (2023). Smart contracts in blockchain technology: a critical review. *Information, 14*(2). https://doi.org/10.3390/info14020117
15. Vaigandla, K. K. (2023). Review on blockchain technology: architecture, characteristics, benefits, algorithms, challenges and applications. *The Mesopotamian Journal of Cybersecurity, 2023*, 73–85. https://doi.org/10.58496/mjcs/2023/012

16. Tahora, S., Saha, B., Sakib, N., Shahriar, H., & Haddad, H. (2023). *Blockchain technology in higher education ecosystem: unraveling the good, bad, and ugly*. arXiv preprint arXiv:2306.04071. [Online]. Available: http://arxiv.org/abs/2306.04071
17. Saha, B., Hasan, M. M., Anjum, N., Tahora, S., Siddika, A., & Shahriar, H. (2023). *Protecting the decentralized future: An exploration of common blockchain attacks and their countermeasures*. arXiv preprint arXiv:2306.11884. [Online]. Available: http://arxiv.org/abs/2306.11884
18. Javed, I. T., Alharbi, F., Bellaj, B., Margaria, T., Crespi, N., & Qureshi, K. N. (2021). Health-id: A blockchain-based decentralized identity management for remote healthcare. *Healthc., 9*(6), 1–21. https://doi.org/10.3390/healthcare9060712
19. Buttar, A. M., Bano, M., Akbar, M. A., Alabrah, A., & Gumaei, A. H. (2023). Toward trustworthy human suspicious activity detection from surveillance videos using deep learning. *Soft Computing, 0123456789*. https://doi.org/10.1007/s00500-023-07971-x
20. Habib, G., Sharma, S., Ibrahim, S., Ahmad, I., Qureshi, S., & Ishfaq, M. (2022). Blockchain technology: Benefits, challenges, applications, and integration of blockchain technology with cloud computing. *Future Internet, 14*(11), 1–22. https://doi.org/10.3390/fi14110341
21. Khalid, M. I., et al. (2023). A comprehensive survey on blockchain-based decentralized storage networks. *IEEE Access, 11*, 10995–11015. https://doi.org/10.1109/ACCESS.2023.3240237
22. Florea, A. I., Anghel, I., & Cioara, T. (2022). A review of blockchain technology applications in ambient assisted living. *Future Internet, 14*(5). https://doi.org/10.3390/fi14050150
23. Djeddai, A., & Khemaissia, R. (2023). PrivyKG: Security and privacy preservation of knowledge graphs using blockchain technology. *Informatica, 47*(5), 137–152. https://doi.org/10.31449/inf.v47i5.4698
24. Bouras, M. A., Lu, Q., Dhelim, S., & Ning, H. (2021). A lightweight blockchain-based iot identity management approach. *Future Internet, 13*(2), 1–14. https://doi.org/10.3390/fi13020024
25. Waseem, M., Adnan Khan, M., Goudarzi, A., Fahad, S., Sajjad, I. A., & Siano, P. (2023). Incorporation of blockchain technology for different smart grid applications: Architecture, prospects, and challenges. *Energies, 16*(2). https://doi.org/10.3390/en16020820
26. Eremina, L., Mamoiko, A., & Aohua, G. (2023). Application of distributed and decentralized technologies in the management of intelligent transport systems. *Journal of Intelligent and Robotic Systems, 3*(2), 149–161. https://doi.org/10.20517/ir.2023.09
27. Du, Z., Jiang, W., Tian, C., Rong, X., & She, Y. (2023). Blockchain-based authentication protocol design from a cloud computing perspective. *Electronics, 12*(9). https://doi.org/10.3390/electronics12092140
28. Mateen, A., & Amir, H. (2016). Enhancement in the effectiveness of requirement change management model for global software development. *Journal of Science International Lahore, 28*(2), 1161–1164. [Online]. Available: http://arxiv.org/abs/1605.00770
29. Schlatt, V., Sedlmeir, J., Traue, J., & Völter, F. (2023). Harmonizing sensitive data exchange and double-spending prevention through blockchain and digital wallets: The case of E-prescription management. *Distributed Ledger Technologies: Research and Practice, 2*(1), 1–31. https://doi.org/10.1145/3571509
30. Wenhua, Z., Qamar, F., Abdali, T. A. N., Hassan, R., Jafri, S. T. A., & Nguyen, Q. N. (2023). Blockchain technology: Security issues, healthcare applications, challenges and future trends. *Electronics, 12*(3). https://doi.org/10.3390/electronics12030546
31. Badirova, A., Dabbaghi, S., Moghaddam, F. F., Wieder, P., & Yahyapour, R. (2023). A survey on identity and access management for cross-domain dynamic users: issues, solutions, and challenges. *IEEE Access, 11*(June), 61660–61679. https://doi.org/10.1109/ACCESS.2023.3279492
32. Mahmood, M. S., & Al Dabagh, N. B. (2023). Blockchain technology and internet of things: Review, challenge and security concern. *International Journal of Electrical and Computer Engineering, 13*(1), 718–735. https://doi.org/10.11591/ijece.v13i1.pp718-735
33. Tan, E., Lerouge, E., Du Caju, J., & Du Seuil, D. (2023). *Verification of education credentials on european blockchain services infrastructure (EBSI): Action research in a cross-border use case between Belgium and Italy* (Vol. 7).
34. Zhu, X., He, D., Bao, Z., Luo, M., & Peng, C. (2023). An efficient decentralized identity management system based on range proof for social networks. *IEEE Open Journal of the Computer Society, 4*(March), 84–96. https://doi.org/10.1109/OJCS.2023.3258188

35. Alanzi, H., & Alkhatib, M. (2022). Towards improving privacy and security of identity management systems using blockchain technology: A systematic review. *Applied Sciences, 12*(23). https://doi.org/10.3390/app122312415
36. Singh, D., Monga, S., Tanwar, S., Hong, W. C., Sharma, R., & He, Y. L. (2023). Adoption of blockchain technology in healthcare: Challenges, solutions, and comparisons. *Applied Sciences, 13*(4). https://doi.org/10.3390/app13042380
37. Rahmani, M. K. I., et al. (2022). Blockchain-based trust management framework for cloud computing-based internet of medical things (IoMT): A systematic review. *Computational Intelligence and Neuroscience, 2022*, 1. https://doi.org/10.1155/2022/9766844
38. Truong, H., et al. (2022). Enabling decentralized and auditable access control for IoT through blockchain and smart contracts. *Security and Communication Networks, 2022*, 1. https://doi.org/10.1155/2022/1828747
39. Mesías-Ruiz, G. A., Pérez-Ortiz, M., Dorado, J., de Castro, A. I., & Peña, J. M. (2023). Boosting precision crop protection towards agriculture 5.0 via machine learning and emerging technologies: A contextual review. *Frontiers in Plant Science, 14*(March), 1–22. https://doi.org/10.3389/fpls.2023.1143326
40. Xiao, Y., et al. (2022). Decentralized spectrum access system: Vision, challenges, and a blockchain solution. *IEEE Wireless Communications, 29*(1), 220–228. https://doi.org/10.1109/MWC.101.2100354
41. Zubaydi, H. D., Varga, P., & Molnár, S. (2023). Leveraging blockchain technology for ensuring security and privacy aspects in internet of things: A systematic literature review. *Sensors, 23*(2). https://doi.org/10.3390/s23020788
42. Friedewald, M., & Kreutzer, M. (2022). *Selbstbestimmung, Privatheit und Datenschutz*. Springer.
43. Sung, C. S., & Park, J. Y. (2021). Understanding of blockchain-based identity management system adoption in the public sector. *Journal of Enterprise Information Management, 34*(5), 1481–1505. https://doi.org/10.1108/JEIM-12-2020-0532
44. Kumar, D., Kumar, S., & Joshi, A. (2023). Assessing the viability of blockchain technology for enhancing court operations. *International Journal of Law, 65*, 425. https://doi.org/10.1108/IJLMA-03-2023-0046
45. Sadique, K. M., Rahmani, R., & Johannesson, P. (2023). DIdM-EIoTD: Distributed identity management for edge Internet of Things (IoT) devices. *Sensors, 23*(8). https://doi.org/10.3390/s23084046
46. Chawla, P., Kumar, A., Nayyar, A., & Naved, M. (2023). *Blockchain, IoT, and AI technologies for supply chain management*. https://doi.org/10.1201/9781003264521.
47. Ghaffari, F., Gilani, K., Bertin, E., & Crespi, N. (2022). Identity and access management using distributed ledger technology: A survey. *International Journal of Network Management, 32*(2), 1–19. https://doi.org/10.1002/nem.2180
48. Wang, Z., Zhang, C., & Mu, X. (2023). Decentralized solution for cold chain logistics combining IoT and blockchain technology. *Journal of network intelligence, 8*(1), 47–61.
49. Uppal, S., Kansekar, B., Mini, S., & Tosh, D. (2023). HealthDote: A blockchain-based model for continuous health monitoring using interplanetary file system. *Healthcare Analytics, 3*(March), 100175. https://doi.org/10.1016/j.health.2023.100175
50. Han, R., Shapiro, G., Gramoli, V., & Xu, X. (2020). On the performance of distributed ledgers for internet of things. *Internet of Things, 10*, 1–11. https://doi.org/10.1016/j.iot.2019.100087
51. Stockburger, L., Kokosioulis, G., Mukkamala, A., Mukkamala, R. R., & Avital, M. (2021). Blockchain-enabled decentralized identity management: The case of self-sovereign identity in public transportation. *Blockchain: Research and Applications, 2*(2), 100014. https://doi.org/10.1016/j.bcra.2021.100014
52. Rathee, T., & Singh, P. (2022). A systematic literature mapping on secure identity management using blockchain technology. *Journal of King Saud University, Computer and Information Sciences, 34*(8), 5782–5796. https://doi.org/10.1016/j.jksuci.2021.03.005
53. Liao, C. H., Guan, X. Q., Cheng, J. H., & Yuan, S. M. (2022). Blockchain-based identity management and access control framework for open banking ecosystem. *Future Generation Computer Systems, 135*, 450–466. https://doi.org/10.1016/j.future.2022.05.015

Chapter 9
Reshaping the Education Sector of Manipur Through Blockchain

Benjamin Kodai Kaje, Ningchuiliu Gangmei, Hrai Dazii Jacob, and Nganingmi Awungshi Shimray

1 Introduction

There may not be much contestation about the scenic beauty of the state of Manipur. Great people of the past have spoken in testimony to this fact. Pandit Jawaharlal Nehru has described Manipur as the 'Land of Jewel', and Lord Irwin called it the 'Switzerland of the East'. Though the landscape is beautiful, it is plagued by various problems. Northeast India is marked by political unpredictability, economic underdevelopment, and cultural marginalisation. Aside from a poor interregional connectivity, the region's development is hampered by a number of intrinsic structural and cultural problems [9]. Widespread corruption has long been a problem in India as a whole. Despite the fact that many academics have emphasised the necessity of good governance in mitigating the problem, they have only seen a linear and symmetrical relationship between governance and corruption [16]. It prevents our country from effectively utilising its resource pool to its greatest potential. For instance, employment, contracts, etc. are awarded to the highest bidder, not necessarily to the most effective [32]. Likewise, law and order, a high cost of transportation, nepotism, interethnic strife, unusual land ownership patterns, insurgency, and corruption are pervasive in Manipur's daily development activities. This has an impact on the construction of crucial infrastructure. Thus, resolving these difficulties might open the door for more robust development [38]. Many of the problems are man-made, and the solution is possible only through the goodwill of the people. Some suggested the lack of development due to years of geographical isolation. Hence, in order to catch up with the rest of India and the world, the state of Manipur needs to welcome

B. K. Kaje (✉) · N. Gangmei
Christ University, Bangalore, India
e-mail: benjamin.kaje@res.christuniveristy.in

H. D. Jacob · N. A. Shimray
OPJS University, Churu, Rajasthan, India

S. M. Idrees, M. Nowostawski (eds.), *Blockchain Transformations*, Signals and Communication Technology, https://doi.org/10.1007/978-3-031-49593-9_9

change and strive forward. One of the best and most promising ways is to adapt to modern and technological change to be at par with the world.

1.1 Brief Background to the Study

Manipur's territory measures 22,327 km^2, or 0.68%, of the nation's total size. According to the 2011 census, Manipur has a population of 2.86 million people, of which 29.20% live in urban areas, and 70.80% do so in rural areas. The geography of the state of Manipur is divided into hills and valleys where the tribals occupy the hills, and the non-tribal Meiteis occupy the valley. The valley consists of about 10% of the total area, whereas the hills constitute about 90% of the total land area.

The month of May 2023 saw an unprecedented ethnic clash between the Meiteis and the Kuki-Zo. There was a prolonged growing tension in the built-up of the conflict. Some of the critical contentions among the ethnic groups is land, job, development, and employment. The Meiteis (non-tribals) felt threatened due to the paucity of land in the state's capital Imphal. The valley holds about 60% of the total population. Since the tribals can buy land from the valley area, on the other hand, the Meiteis cannot purchase any land from the tribals, thereby causing a certain amount of disquiet among the Meiteis. The tribals in Manipur are of the opinion that all the developmental works are generally concentrated in the valleys of Imphal. There has been constant uneasiness on the part of the tribals, both the Nagas and Kukis, that the government had focussed primarily on the Imphal valley.

Having apprised, in a nutshell, of the situation in Manipur, blockchain technology can be a bridge between the Meities and the tribals of Nagas and Kukis. Blockchain technology could bring about parity to a large extent, thereby avoiding certain social unrest and disturbances in Manipur. Through the use of technology, it can narrow the various developmental gaps created with time. Gilder, 1994, as cited in Jae Park [25], the most fundamental paradigm change in digital communication and educational technologies may well be brought about by blockchain technology. There has been much debate and deliberations about the environmental hazards caused by modernisation. The need of the hour is to reduce, if not stop, ecological damages caused by various developmental activities, including the education sector. The use of blockchain technology will aid in navigating a sustainable, environment-friendly future and using green energy. One can remotely say that education is the cause of the industrial revolution, which, in a way, hugely impacted the environment. Along with the developmental growth, it caused much damage to nature. Blockchain technology should take the lead to change the future for the better. It could change the future for the betterment of all in the state.

The chapter will be mainly divided into two parts. The first part will begin by briefly analysing what blockchain is. Thereafter, it will explore how blockchain could be profitably used in the teaching-learning experience of the student-teacher relationship. The second part will delve into the area of the use of blockchain in the administrative sector. If taken well by the government, this chapter could be a trailblazer to the use of technology in the transformation of the educational sector of

Manipur. Finally, a discussion is opened as to the feasibility of blockchain in Manipur with all its challenges and hurdles. Blockchain may not solve all the societal ills and environmental challenges, but at the same time, it will provide a good platform to launch a secure, eco-friendly, and sustainable future. There is a short recommendation that ends with the need to introduce blockchain in Manipur for the good of all its citizens.

1.1.1 What Is a Blockchain?

Blockchain is a rather unwonted word, rarely used in ordinary parlance. Blockchain technology can get rid of intermediaries and enable direct transactions and verification [24]. The Internet and blockchain are both pieces of technology that allow for the exchange of money and other goods and services digitally. Blockchain is based on a principle known as Decentralized Ledger Technology (DLT). Blockchain is the name given to a wide range of technologies that offer immutable ledgers that are replicated and synchronised across many nodes. The additions to the ledger are confirmed by network consensus using a special method, guaranteeing data consistency. The key innovation or advantage of blockchains is their ability to reach consensus and uphold data consistency even in the face of a small number of malevolent nodes. With the security and dependability of the ledger data still intact, this enables the system to operate in settings where there is some level of trust. Public permissionless blockchains are also devoid of central administrators and do not demand the confidence of a third party [29].

Satoshi Nakamoto is credited with developing the first application of current blockchain technology. It enables a network to check the transaction history of an electronic coin that a user submits for payment and confirm that the coin has not already been spent [26]. To sum up, a blockchain, in its most basic form, is a kind of distributed ledger technology (DLT) in which transactions are recorded with an irreversible cryptographic signature. A distributed ledger, often known as a blockchain, is a safe, digital, and decentralised record of all transactions. The administration of education, learning, and training at all levels should consider trust [13]. It is a digital progression inducted into various aspects of the social milieu to bring about improvement, efficiency, and accuracy in offices and places of work. It is a part of the scientific contribution to society to uplift and upgrade the standard of living for the masses.

2 Literature Review

2.1 Role of Blockchain in Teaching-Learning

Over the years, technology has come to the aid of humanity in bringing about efficiency and better coordination in various aspects. Education is no exception to it. Blockchain technology shows tremendous promise for students and teachers in its extensive use for designing and implementing learning activities, conducting formative evaluations, and tracking the entire learning process [4]. Moreover, the students, teachers, parents, and institutions of higher learning share the learning process and its results. It enables learning progress to be openly shared between schools, instructors, and parents, which minimises the national education administrative department's involvement in students' learning processes and the evaluation of results, fostering educational equity and enhancing management effectiveness [37]. This technology facilitates safely transferring their learning records from one institution to another. A foundation of learning logs supports it. As a smart contracting technology, blockchain has advanced due to its immutability, provenance, and peer execution, which can provide e-learning with new degrees of security, trust, and transparency [27].

The main value-adding characteristics of blockchain technology for e-learning are the peer execution and consensus process for assessments and the distributed, immutable storing of learner records. Transparent smart contracts can be run on numerous blockchain peers to assure the impartiality of processes, thereby enhancing the system's reliability. It is beneficial when creating credentials and tests. Although consensus is not strictly required for curriculum personalisation, credential production is supported by the distributed storage of such data [21]. Children in various learning groups and individual students working with instructors can access TeachMePlease's (TMP) and successLife online and offline using blockchain. It provides information about educational programs, educational sources, and video archives of seminars, courses, and events. On its portal, SuccessLife has also posted information on events, including coaching, webinars, videos, and workshop assignments [30]. Blockchain-powered tokens have the potential to significantly improve the motivation and engagement of students within non-formal and informal educational platforms like online courses and MOOCs, as evidenced by the numerous private initiatives aimed at improving the effectiveness and engagement of learning and teaching processes for learners and content providers [33].

2.1.1 Blockchain in the Educational Administrative Sector

In the modern world, certificates are a crucial tool for demonstrating success in lifelong learning, but they are also easily forged. Blockchain assists in resolving this issue. Data storage, formative assessment, data security, a universal database, smart contracts, and many other things are among the clear advantages of blockchain [2]. Blockchain technology can successfully stop academic and learning fraud and maintain the level of security required for the certificate to be universally recognised by institutions and companies throughout the globe. Although it is a relatively new field, digitally preserving student credentials can reduce paper and printing costs and avoid losing or damaging papers if stored manually [5]. Its features provide a new set of opportunities to improve the security, trust, and effective use of academic material and to issue, trade, exploit, and verify it while securely facilitating the development of new use cases [10]. The main benefit of blockchain is that it will be used to issue digital certifications, which will use a public Blockchain to record the digital signatures related to such certifications. Not only would a certificate's proof of validity be kept on a blockchain, but the certificate itself would also be kept there, making it eternal and unchangeable [14].

The learners have trouble validating their older, further away higher education degrees. If the college where the learner obtained their degree no longer exists, individuals who still pass through it can distrust the same certifications. Such circumstances cannot occur if the records are stored on a blockchain system since the records are held in numerous ledgers that are all preserved individually [34]. Finally, blockchain helps authenticate educational institutions' credentials, and students could be a significant step toward overcoming the challenge of reaching the unreachable in less developed countries and remote regions. However, access to better infrastructure remains a pending issue, especially regarding Internet access [20]. A blockchain is a new platform for recording learning successes beyond transcripts and certificates, particularly in how learning or teachings were done and achieved by retaining digital hashes of learning activities using smart contracts [27].

2.1.2 Role of Education Towards a Corrupt-Free Society

The basic role of education in society is twofold; to preserve the good aspects of society and to provide necessary changes that are harmful to society. Though modern education reached Manipur over a century back, corrupt practices too have grown among all sections and various departments in the government. Value-based education needs to be inculcated in young minds so that they become upright citizens who are guided by principles. Neglecting ethical principles, which ought to be the foundation of all worthwhile education, has resulted in ineffective, decadent, and pointless learning [6]. Teachers need to be convinced of their roles to form good citizens so that the future of the state is in safe hands.

In Manipur, the term 'corruption' primarily refers to the misuse of public funds by those in positions of authority, those who hold official positions in administrations and public offices, and their cohorts, as well as the practice of selling or buying government jobs in exchange for money or personal favour. Competition in politics has been linked to transparent government operations, an engaged media, and an informed civil society. Because of these controls, political and administrative corruption is the exception rather than the rule [1]. The fraud and corruption, as well as systematic errors in resource allocation, made the poor parents of Manipur to send their children to private schools rather than to government schools [28].

In the case of Manipur, corruption and red tape frequently have a negative impact on the excitement and successful completion of projects by SMCs (School Management Committee). There are various levels of corrupt practices in the offices. From high-ranking officials, it is percolated down to the lowest clerk in the office. Unfortunately, what is put on paper doesn't always reflect reality. Many SMC members are caught in this predicament, which contributes to an unintentional lack of openness [18]. Many educated young people leave Manipur in quest of employment when there aren't enough job possibilities there. Public infrastructure, including roads and buildings, is built with defective materials, and some of it even just exists on paper. As a result, the development of a black economy and the backwardness of the infrastructure have been brought about by decades of insurgency and pervasive corruption [17].

Additionally, corruption, nepotism, bribery, favouritism, and connections to influential politicians and bureaucrats are rampant throughout the hiring of teachers. The fact that many teachers would improperly hire a proxy or replacement teacher in the next village wherever they are posted is another significant issue with government schools in Manipur. Government schools, particularly those in rural areas, are closing as a result of this sort of bad conduct [3]. In Manipur, it is still extremely difficult to integrate reforms to public policy and institutional structures that will effectively reduce corruption and consolidate good governance. Growing sensitivity to the public or a decline in the moral standards of all government players, or both, led to an increase in awareness of the issues related to corruption. Local institutions like the police, government, judiciary, and legislative are heavily entwined in the web of organised crime and corruption. The state political system could no longer heal on its own. Unfortunately, there is no organised effort or campaign in the state to combat the corruption threat [31].

2.2 Findings of Blockchain in the Education Sector

Preeti Bhaskar, et al., did a systematic literature review in which they found that blockchain in education management is a young discipline. The result of the analysis demonstrates that while blockchain technology in education is still a relatively new field, it has a lot to offer the entire field of education. Additionally, direct verification and transaction are made possible by eliminating third parties as intermediaries [7]. Blockchain technology is dependable, and a decentralised network allows for the constant updating of all transaction records databases. The network is unaffected by the breakdown, guaranteeing the remarkable resilience of apps built using blockchain technology. It operates on a very secure network of effective, tamper-proofed nodes that follow pre-established procedures [8]. Strong cybersecurity capabilities of blockchain technology frameworks have been applied in numerous industries. The use of blockchain technology in education is still in its infancy. Additionally, the blockchain is a mechanism for issuing, confirming, and exchanging certifications in educational environments [12].

Blockchain enables the management and preservation of academic credentials, including transcripts, certificates, academic records, degrees, etc. Its features and applications in the educational field can be fully utilised to enhance the entire teaching-learning process, provide fair evaluation of both students and teachers, improve performance, motivate both students and teachers by offering them rewards, manage records, detect fraud, and other things [35]. It can be utilised across several platforms, allowing it to be employed from the beginning of the learning process with e-learning, registration, successes, and values that are well recorded and displayed to serve finances, including tuition payments, libraries, and other educational needs required. It is anticipated that education will keep raising the bar for educational standards by offering high-quality technical advancements that are put to use as needed [23].

Education professionals can save student learning achievements, academic certifications, credit management, etc. using blockchain-based solutions. Blockchain can be used to carry the study article publication in a timely manner [36]. Moreover, the construction of a single educational environment, the development of network communities, the interchange of technology and scientific knowledge, and the copyright protection of network participants are the most significant benefits of educational blockchain technologies [11]. Additionally, almost all of education's subdomains, including financial acquisition, school funding, donations, salary, teacher professional development, human resource management, cross-border academic credit recognition, certification and degree transfers, monitoring student learning, attesting actual abilities, and financial blockchain can be used in education [22, 25].

2.2.1 Hurdles Towards Blockchain Technology

Given the present condition of Manipur, it would be a true challenge towards the implementation of blockchain technology. There are issues of power supply to all parts of Manipur. There are issues related to the network, and some of the terrains in Manipur are hard to reach. The higher education stakeholders now appear to be less aware of the social benefits and educational/instructional potential of blockchain technology, despite all of its potential and advantages [19]. There are ample benefits, yet, people are to realise the importance and potential gains even in the field of education through blockchain. Implementing this new technology into the administrative sector would be a big challenge for the government. There are issues relating to the implementation of the technology in the office. There could be serious doubt about the technological know-how of the administrative sector employees in the government offices' education department. The average human tendency is that people generally have an aversion to change. Hence, to ask people in the offices to learn computers and new technologies would be a tough call on the part of the government.

Investing in the blockchain in educational institutions would be a big ask, especially in private institutions where viability could be an issue. Many educational institutions in Manipur barely manage to pay the teachers in schools. The challenge remains how technology can enhance the teaching-learning method if made into use for the same. Besides, there is the issue of insurgency which has crept into the social milieu of the state for some time. Any developmental works in the state have to encounter this challenge of unwanted elements in society. The common populace of the state needs to cooperate with the state machinery in order to bring about change in society. If all the sections of society collaborate with the state government, change is possible for good. There should be a concerted effort from all fronts of society to bring about that change in society.

2.2.2 Future of Blockchain Technology in Manipur

The role of education towards bringing development in Manipur is an undeniable fact. Within a span of about the last hundred years, there has been tremendous growth in different aspects of society. There has been improvement in the health sector, transportation, communications, better livelihoods, and even accessible educational institutions. Yet there are lots to be taken into consideration. Though the state of Manipur is small, there has been rampant corruption prevalent in the state. The people of Manipur will bear testimony to this fact. Applying the participant observation method, if blockchain is put into use, many corrupt practices can be away with.

Social unrest has been brewing in the minds, especially of the tribals, due to a lack of accessibility in the administrative sector. If blockchain is put into use, digital bridging can help to bring about change overall in the state. The study reveals that while blockchain technology in education is still a relatively new field, it has a lot of potential to benefit the wider education sector [24]. In order to bring about education both in the classroom and in the administrative sector, blockchain technology is the need of the hour for a state like Manipur. It is a known fact that being part of the state, Manipur is stiff into corruption. Blockchain technology can bring about much-needed change in the administrative sector. Since there will be transparency and high security in dealing with and issuing certificates, people across all walks of life will appreciate the effort to imbibe and inculcate this new technology.

Even though the state's literacy rate is higher than the national rate, the quality of education is still low. Hundreds of students go to other states to get higher education as the state cannot meet the learners' needs due to a lack of infrastructure. Furthermore, the state frequently experiences bandhs and blockades, which seriously impair the ability of education to function. Conflicts may arise as a result of the state's vulnerable situation. With its distributed ledgers, the education blockchain would establish innovative standards for crypto-learning and crypto-administration that are recognised by all organisations and countries, improving the objectivity, validity, and control of information without being harmed by socioeconomic instability [25].

3 Discussion

The task ahead is challenging for the government and the educational institutions. With the given limited access to technology, and to adopt and feed the latest technology is a herculean task. It is not impossible, but it calls for lots of political will and dogged determination on the part of private educational institutions. Education is anticipated to strengthen the country and drive out harmful social habits like corruption. The prevalence of corruption in the educational sector is regrettable. Teachers are responsible for guiding their charges toward the light, but they cannot do so if they are still navigating the shadows. Without rooting out corruption in the state, various ideals and innovations would be of less use. Investment in blockchain will be worthwhile because it can be game-changing in the administrative sector as well as in educational institutions. Transparency and efficiency will become the new order of the day, provided the government and educational institutions are willing to cash in the best of blockchain technology. According to the study, practitioners, economists, and computer scientists can work together to successfully teach blockchain technology [15].

Most modern technologies do have some loopholes, and blockchain is no exception to it. As much as there are advantages to various technologies, there could be equal risks in their use as well. The issue of privacy and threat exists in the blockchain. It appears that the EU views blockchain as a potential risk to citizens' rights to and obligations for their data ownership [33]. This issue needs to be addressed properly so that individual data does not fall into the wrong hands. The safety and security of each individual is paramount. Besides, there are murmurs that blockchain technology is slow. Since it is in its initial stage, the development of any technology takes time to perform at the optimum level. It is the common aspiration of all that blockchain technology will help to improve life in all walks of life.

Given the quick pace of growth and advancement of science and technology, it is bound to improve. If blockchain is inculcated into administrative and educational institutions, lots of unwanted printing could be avoided. Even the age-old practice of writing examinations could have a re-look and some other forms of giving examinations be adopted. Digital examinations could be a fitting alternative, as it is accurate, prompt, and error-free and have no room for corruption. If printings are reduced, simultaneously there would be less felling of trees. This will improve the ecology and the overall ecosystem. One of the paramount duties of education today is the preservation of ecology or the ecosystem. Therefore, if blockchain is put to good use, it will improve the social fabric, bridging the gap between the tribals and non-tribals in the context of Manipur.

Over and above, if blockchain technology is introduced into the administrative sector and in educational institutions, it could create job chances for the youths in Manipur. Thousands of students move outside the state to Delhi, Mumbai, Chennai, Bangalore, and other cities in search of better job opportunities. If the government of Manipur can create such an avenue for the youth, then the youths would remain in the state and the economy of the state would begin to improve gradually. Many youths are media savvy, but for lack of opportunity, their talents go wasted and buried. It is high time that those in authority should think about the future of tomorrow.

4 Recommendation

There are many students from the less-privileged groups in society who cannot afford the prestigious schools in the towns and cities. Technology should come to the aid of these students so that students are left behind due to a lack of opportunities. It would be the prerogative of the government to act as a mediator for low-cost but quality education through the use of modern technology. Despite being aware of the benefits of technology in education, people do not take it seriously or think about changing how education is delivered. The state's educational system must be modified in several ways, starting with funding for infrastructural renovations, other educational upgrades, and teacher recruiting. Even if some schools have all the required resources, they still don't meet expectations. Regular teacher training sessions must be done to stay current on new information and encourage the use of technology in teaching and learning.

The state government of Manipur needs to give serious thought to adopting blockchain into educational institutions and more, especially into the administrative sector. Blockchain might take time and probably be expensive, yet it will bring about efficiency, competence, and transparency in the administration. There are apprehensions that blockchain technology is tedious and slow due to a variety of steps or blocks to be followed, yet with time, and it is bound to improve its efficiency. Although this investigation also suggests some potential uses of blockchain technology in different components of the training framework, more applications can be introduced into the educational framework to fully utilise blockchain technology. This examination provides an infrastructure for educational organisations, strategy creators, and specialists to investigate various areas where blockchain technology can be implemented [24]. A detailed study could be encouraged as to the best way to implement this new technology in Manipur. It could be expensive, the power supply could be an issue, and internet connectivity could upset plans in rural areas. Whatever may be the challenges, if the people have the will and the authority have the goodwill for the people, all things are doable and achievable.

5 Conclusion

The future belongs to the world of technology. Education needs technology to grow and progress. The time is ripe to introduce blockchain technology in various aspects, beginning with the educational administrative sector as well as in the teaching-learning ambience. In order to cope with the standard and competition of the world, Manipur needs blockchain technology to give a fillip to its education in the teaching-learning ecology and also in its administrative set-up. Transparency, immutability, and easy accessibility can be seen, and bottleneck administration in the education sector can be avoided. Blockchain is in its early years, and its ultimate outcome is yet to be fully explored and discovered. It has immense potential to render in all spheres of life and, more significantly, in the educational sector. The government needs to care and make technology accessible and affordable for the less privileged and those people of lower economic strata. The present generation of the young has to be given good and value-based education with the best of modern technologies. When the youths are formed with principles in life, there will be less corruption and quality education will be possible.

The desire of every citizen of Manipur and India would be to see a robust and vibrant state and country. For a country and state to grow, corrupt practices will consume the good works of modern technology. Hence, while desiring a developed state, there should be honest, value-based citizens yearning for the overall growth of the state. Blockchain technology should facilitate a quicker and faster pace of developing and bringing about change in a state like Manipur. With the introduction of blockchain in administrative and educational institutions, there will be all-around progress in the state. Besides, with less printing due to digital certification and verification, it will remotely augur well in the environmental sector. If all the states and the country as a whole adopt this technology, it will reduce corruption and improve the ecosystem in the long run.

If one delves deep into blockchain technology's diverse, multi-faceted utility, it can offer in many areas of society. It may not be able to erase corruption entirely from the administrative sector of the government, nor will educational institutions go paperless. Nevertheless, the effort put in, however small and insignificant, will go a long way in bringing about transparency and efficiency in public offices. If the use of paper printing can be reduced to some extent, then the purpose of blockchain technology has partially achieved some of its objectives in the context of Manipur. If education does not help in bringing about a sustainable future, the very purpose of education is defeated. Education should help to perverse nature and erase the social evils like corruption, casteism, tribalism, and religious fanaticism from society.

References

1. Akoijam, A. (2012). *Making sense of corruption in Manipur*. Kangla Online. https://kanglaonline.com/2012/09/making-sense-of-corruption-in-manipur/
2. Arndt, T. (2019). An overview of blockchain for higher education. *IC3K 2019 – Proceedings of the 11th International Joint Conference on Knowledge Discovery, Knowledge Engineering and Knowledge Management, 3*, 231–235. https://doi.org/10.5220/0008343902310235
3. Aveivey, D. (2019). An analysis on education of Poumai Naga tribe in Manipur. *International Journal of Humanities and Social Science Invention (IJHSSI), 8*(1), 18–24. http://www.ijhssi.org/papers/vol8(1)/Version-1/C0801011824.pdf
4. Awaji, B., Solaiman, E., & Albshri, A. (2020). *Blockchain-based applications in higher education: A systematic mapping study* (pp. 96–104). ACM International Conference Proceeding Series, July. https://doi.org/10.1145/3411681.3411688
5. Bagchi, S. (2019, December). *AI, blockchain to reshape education sector in 2020*. Cxotoday.Com. https://www.cxotoday.com/news-analysis/ai-blockchain-to-reshape-education-sector-in-2020/
6. Balaji Iyer, D. R. (2013). Value-based education: Professional development vital towards effective integration. *IOSR Journal of Research & Method in Education (IOSRJRME), 1*(1), 17–20. https://doi.org/10.9790/7388-0111720
7. Bhaskar, P., Tiwari, C. K., & Joshi, A. (2020). Blockchain in education management: Present and future applications. *Interactive Technology and Smart Education, 18*(1), 1–17. https://doi.org/10.1108/ITSE-07-2020-0102
8. Casey, M., Crane, J., Gensler, G., Johnson, S., & Narula, N. (2018). The impact of blockchain technology on finance: A catalyst for change. In *Geneva Reports on the World Economy*. https://cepr.org/publications/books-and-reports/geneva-21-impact-blockchain-technology-finance-catalyst-change

9. Das, P. (2022). Challenges to the development of the northeast through the act east policy. *Strategic Analysis, 46*(5), 473–493. https://doi.org/10.1080/09700161.2022.2119698
10. Delgado-Von-eitzen, C., Anido-Rifón, L., & Fernández-Iglesias, M. J. (2021). Blockchain applications in education: A systematic literature review. *Applied Sciences (Switzerland), 11*. https://doi.org/10.3390/app112411811
11. Fedorova, E. P., & Skobleva, E. I. (2020). Application of blockchain technology in higher education. *European Journal of Contemporary Education, 9*(3), 552–571. https://doi.org/10.13187/ejced.2020.3.552
12. Gräther, W., Augustin, S., Schütte, J., Kolvenbach, S., Augustin, S., Ruland, R., Augustin, S., & Wendland, F. (2018). *Blockchain for education: Lifelong learning passport*. https://doi.org/10.18420/blockchain2018.
13. Grech, A., Balaji, V., & Miao, F. (2022). Education and blockchain. In *UNESCO* (Issue December). UNESCO. https://doi.org/10.56059/11599/4131
14. Grech, A., & Camilleri, A. (2023). Blockchain in education. In A. I. dos Santos (Ed.), *Inamorato dos Santos, A*. https://doi.org/10.2760/60649
15. Gutowski, P., Markiewicz, J., Niedzielski, P., & Klein, M. (2022). Blockchain in education: The best teaching models. *European Research Studies Journal, XXV*(4), 253–266.
16. Han, J. (2023). How does governance affect the control of corruption in India? A configurational investigation with Fs/QCA. *Economies, 11*(2). https://doi.org/10.3390/economies11020043
17. Haokip, T. (2019). *Deepening crisis in Manipur*. The Statesman. https://www.thestatesman.com/supplements/north/deepening-crisis-manipur-1502769135.html
18. Heigrujam, R., & Philomina, M. J. (2021). School management committee in elementary schools of Imphal west district: A brief review. *Journal of Research & Method in Education, 11*(2), 1–12. https://doi.org/10.9790/7388-1102010112
19. Hussain, I., & Cakir, O. (2020). Blockchain technology in higher education: Prospects, issues, and challenges. In S. Mahankali & S. Chaudhary (Eds.), *Blockchain in education: A comprehensive approach – Utility, use cases, and implementation in a university* (pp. 1–293). IGI Global. https://doi.org/10.4018/978-1-5225-9478-9.ch014
20. Kwok, A. O. J., & Treiblmaier, H. (2022). No one left behind in education: Blockchain-based transformation and its potential for social inclusion. *Asia Pacific Education Review, 23*(3), 445–455. https://doi.org/10.1007/s12564-021-09735-4
21. Lam, T. Y., & Dongol, B. (2022). A blockchain-enabled e-learning platform. *Interactive Learning Environments, 30*(7), 1229–1251. https://doi.org/10.1080/10494820.2020.1716022
22. Loukil, F., Abed, M., & Boukadi, K. (2021). Blockchain adoption in education: A systematic literature review. *Education and Information Technologies, 26*(5), 5779–5797. https://doi.org/10.1007/s10639-021-10481-8
23. Lutfiani, N., Aini, Q., Rahardja, U., Wijayanti, L., Nabila, E. A., & Ali, M. I. (2021). Transformation of blockchain and opportunities for education 4.0. *International Journal of Education and Learning, 3*(3), 222–231. https://doi.org/10.31763/ijele.v3i3.283
24. Maulani, G., Musu, E. W., Soetikno, Y. J. W., & Aisa, S. (2021). Education management using blockchain as future application innovation. *IAIC Transactions on Sustainable Digital Innovation (ITSDI), 3*(1), 60–65. https://doi.org/10.34306/itsdi.v3i1.525
25. Park, J. (2021). Promises and challenges of blockchain in education. *Smart Learning Environments, 8*(1), 1–13. https://doi.org/10.1186/s40561-021-00179-2
26. Popovski, L., & Soussou, G. (2018). A brief history of blockchain. In *ALM pubication* (pp. 1–20). Legaltech News.
27. Raimundo, R., & Rosário, A. (2021). Blockchain system in the higher education. *European Journal of Investigation in Health, Psychology and Education, 11*(1), 276–293. https://doi.org/10.3390/ejihpe11010021
28. Salam, J. (2014). Theft, corruption, and parental school choice in Manipur. *Economic and Political Weekly, 49*(12), 22–24.

29. Satybaldy, A., Hasselgren, A., & Nowostawski, M. (2022). Decentralized identity management for E-health applications: State-of-the-art and guidance for future work. *Blockchain in Healthcare Today, 5*(Special issue), 1–10. https://doi.org/10.30953/bhty.v5.195
30. Sharma, S., & Batth, R. S. (2020). *Blockchain technology for higher education sytem: A mirror review* (pp. 348–353). Proceedings of International Conference on Intelligent Engineering and Management, ICIEM 2020, October. https://doi.org/10.1109/ICIEM48762.2020.9160274
31. Singh, S. (2022). *Mayhem of corruption in Manipur*. Imphal Free Press. https://www.ifp.co.in/opinion/mayhem-of-corruption-in-manipur
32. Sophia, A. (2020). Corruption: A malaise in need of a remedy. *Technium Social Sciences Journal, 12*, 126–136. https://doi.org/10.47577/tssj.v12i1.1734
33. Steiu, M.-F. (2017). Blockchain in education: Opportunities, applications, and challenges. *First Monday, 25*(9), 1–14.
34. Subramanian Iyer, S., Seetharaman, A., & Ranjan, B. (2021). Researching blockchain technology and its usefulness in higher education. *Computer Science & Information Technology (CS & IT), 27*, 27–48. https://doi.org/10.5121/csit.2021.111203
35. Supriadi, A., Iqbal, M., Pratista, A., Sriyono, D., & Buanasari, D. (2022). Blockchain and IoT technology in transformation of education sector. *Frontier Technology, 2*(2), 44–53. https://doi.org/10.34306/bfront.v2i2.208
36. Vishnu, S., Raghavan Sathyan, A., Susan Sam, A., Radhakrishnan, A., Olaparambil Ragavan, S., Vattam Kandathil, J., & Funk, C. (2022). Digital competence of higher education learners in the context of COVID-19 triggered online learning. *Social Sciences & Humanities Open, 6*(1), 100320. https://doi.org/10.1016/j.ssaho.2022.100320
37. Zhou, L., Lu, R., & Wang, J. (2020). Development status, trends and challenges in the field of blockchain and education. *Journal of Physics: Conference Series, 1621*(1), 012112. https://doi.org/10.1088/1742-6596/1621/1/012112
38. Ziipao, R. (2020). Deepening critical infrastructures in Northeast India: People's perspective and policy implications. *Strategic Analysis, 44*(3), 208–223. https://doi.org/10.1080/09700161.2020.1787686

Chapter 10
Exploring the Intersection of Entrepreneurship and Blockchain Technology: A Research Landscape Through R Studio and VOSviewer

Nisha Kumari, Bangar Raju Indukuri, and Prajeet Ganti

1 Introduction

Due to the decentralized nature of blockchain, there is no need for an intermediary organization. It is a cloud database system that keeps a list of data entries that is constantly expanding and is validated by the participating nodes. The data, which is maintained in a public ledger, contains specifics about every transaction ever made [1]. It offers a distributed, transparent, secure, auditable, and immutable ledger [2]. Blockchains also make it possible for peer-to-peer (P2P) networks to automatically do smart contracts [3]. With its attributes of decentralization, immutability, collaborative maintenance, traceability, openness, and transparency, blockchain technology can achieve the decentralized preservation of actual, effective, and authentic data at a minimal cost [4].

In 2008 when the blockchain concept was integrated with several other technologies and computer ideas, modern cryptocurrencies were created as electronic payment methods protected by cryptographic procedures rather than a centralized database or authority [5]. Despite the initial hype around cryptocurrencies, the potential of blockchain applications extends far beyond Bitcoin. Blockchain became a top strategic priority for nations and entrepreneurs globally, and they began to act accordingly [6]. More specifically, more than 50% of CEOs in various nations globally reported that they ranked blockchain as one of their companies' top five priorities, according to Deloitte [7].

The term "blockchain technology" refers to a new trending group of digital innovations that could fundamentally alter entrepreneurial ecosystems, particularly

N. Kumari (✉) · B. R. Indukuri · P. Ganti
Department of Entrepreneurship, GITAM School of Business, GITAM Deemed to be University, Visakhapatnam, Andhra Pradesh, India
e-mail: nkumari4@gitam.in

S. M. Idrees, M. Nowostawski (eds.), *Blockchain Transformations*, Signals and Communication Technology, https://doi.org/10.1007/978-3-031-49593-9_10

those that are currently vulnerable [8]. Different sorts of spatial affordances—potentialities derived from proximity to something—are provided by entrepreneurial ecosystems, which explicates why some produce better results than others [9]. Therefore, digital technology may serve to facilitate the process of identifying and seizing entrepreneurial opportunities as well as help some entrepreneurial ecosystems overcome specific shortcomings in spatial affordances [10].

As a result, the general idea is consistent with the entrepreneurial potential of blockchain technology. Because technology makes it easier to take advantage of business possibilities and overcome obstacles, using blockchain technology successfully as a tool can advance entrepreneurship. The study uses bibliometric analysis to concentrate on "entrepreneurship"and "blockchain technology" in this context.

Bibliometric analysis is a well-known and reliable approach for analyzing and interpreting vast amounts of scientific data. It enables us to illuminate the borders of a field while delving into the details of its evolutionary history [11]. Descriptive analysis and content analysis are the two main methodologies used in bibliometric analysis [12]. Researchers conduct a descriptive analysis to provide data that practitioners and researchers can use as a guide to follow the development of research and anticipate probable future trends [13]. By focusing on the sources, themselves instead of citation- and author-level analyses, the content analysis allowed academics to gain thorough conceptual insights [14].

To fully evaluate the available blockchain and entrepreneurial publications between 2017 and May 2023, this research will combine the benefits of qualitative assessments with computer technology. To be more specific, we assessed publication performance and identified the conceptual, intellectual, and social structure of the blockchain and entrepreneurship current literature using the bibliometrix R-package and VOS viewer software.

Along with enhancing the reliability and transparency of the analysis, the researcher also tries to add the following. In this study, we first carry out descriptive analysis to demonstrate the evolution of blockchain technology in entrepreneurship research's popularity and evaluate the value of publications using a variety of metrics (e.g., the year of the first publication, the number of publications, citations, h-index). This makes it possible to get organized data for studies on blockchain technology in entrepreneurship. Further, researchers explore co-occurrence, thematic map, co-citation, and collaboration worldmap using content analysis.

2 Methodology

The bibliometric analysis involves extracting documents from a single database and processing them statistically and qualitatively [15]. With reference to the findings of international studies on entrepreneurship and the relationship and influence of blockchain technology, papers in the present research on blockchain technology and entrepreneurship were retrieved using the Web of Science. The researchers made use of the Web of Science (WOS), widely regarded as the best method for producing citation data for research evaluation. Numerous global citation studies have used WOS, which includes the three ISI citation databases, as their foundation [16]. To examine the perspectives of all researchers in the field, the current study examines all types of publications in addition to articles. The subject filtration was used, though, to concentrate on blockchain technology in relation to entrepreneurship.

For bibliometric analysis, the researcher adhered to the following procedures (Fig. 10.1). We searched across "titles, abstracts, and keywords" in the Web of Science database for the keywords "blockchain" AND "entrepreneurship" to collect bibliometric metadata concerning blockchain technology and entrepreneurship.

Initially, 382 documents were retrieved when the researcher first utilized the aforementioned string. Additional measures were performed for inclusion and exclusion to extract the most important documents that demonstrate the potential of blockchain in entrepreneurship. To start, based on relevancy, the search was first limited to Business, Management, Economics, Business Finance, Operations Research Management, and Telecommunications. As other areas represent information related to law, mathematics applied, development studies, etc. The database found 236 documents by focusing the search on the Business, Management, Economics, Business Finance, Operations Research Management, and Telecommunications subject areas. To comprehend the emergence and development of the topic over time, the researcher did not do year bases exclusion because blockchain technology in terms of entrepreneurship is a relatively new concept.

Descriptive statistics, such as the average number of publications each year, the most productive author, institutions, and nations, as well as the most often cited author, were computed using the Bibliometrix R package. Content analysis was performed to assess the final dataset's co-occurrence, theme map, co-citation, and cooperation worldmap using the bibliometrix R package and VOSviewer simultaneously.

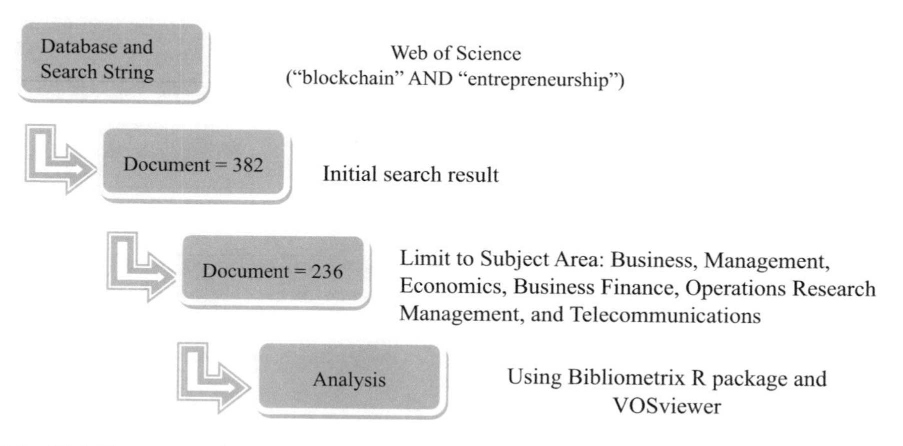

Fig. 10.1 Data processing steps

3 Results

3.1 Descriptive Analysis

3.1.1 Main Information About Data

Major findings using R studio from the investigation, including pertinent information on documents, keywords, nations, and authors illustrated in Table 10.1. The authorship provides extensive and valuable data on the status of the authors and the authors' collaboration [11]. According to Table 10.1, the 236 documents in the sample were published in 121 sources and were written by 621 writers who are affiliated with 422 affiliations in 60 different countries or regions.

Table 10.1 Summary of data

Descriptions	Results
Main information about data	
Sources (Journals, Books, etc.)	121
Documents	236
Annual growth rate %	23.63
Document average age	2.25
Average citations per doc	17.01
References	13,221
Document contents	
Keyword plus (ID)	490
Author's keywords (DE)	910
Authors	
Authors	621
Authors of single-authored docs	33
Authors collaboration	
Single-authored docs	37
Co-authors per doc	3.07
International co-authorship %	35.59
Affiliations	331
Countries	47
Publications	187
Document type	
Articles	143
Articles (early access)	21
Articles (proceeding papers)	1
Editorial material	4
Proceeding papers	49
Review	14
Review (early access)	4

3.1.2 Number of Publications per Year

The first study on blockchain, titled "Blockchain- properties and misconceptions," by Daniel Conte de Leon was published in the Web of Science databases in December 2017. It addressed the common misconceptions about the characteristics of blockchain technologies as well as the difficulties and potential solutions for the development and use of distributed ledger technology and system.

Figure 10.2 depicts the 237 documents included in the sample's annual number of publications and citations. As shown in the figure, there is no chronological incline or decline in publication. The publication in reference to blockchain technology and entrepreneurship has been started in 2017 and most publications were made in 2022, that is, 64. Based on the production and annual growth rate, that is, 23.63%, the researchers can say the topic is having a lot of scope in research. Additionally, the average age of the paper is 2.25 and there are 17.01 citations on average for each document, showing that new research is constantly conducted and novel theories and concepts are continually being explored.

3.1.3 Most Relevant and Citated Journals

This study found 237 publications from 121 sources. The Hirsch index (h-index) of each publication is used to find the most prominent journals in blockchain technology and entrepreneurship research. The H-index is the number of articles by an author or journal that have been cited at least h times in other works. It is widely used to assess an author's or journal's research performance. The top 20 journals are selected based on h-index (see Table 10.2). The number of publications (NP), the total number of citations (TC), and the year of the first publication (PY-start) are also all provided. Researchers can consider these top 20 articles to be the most

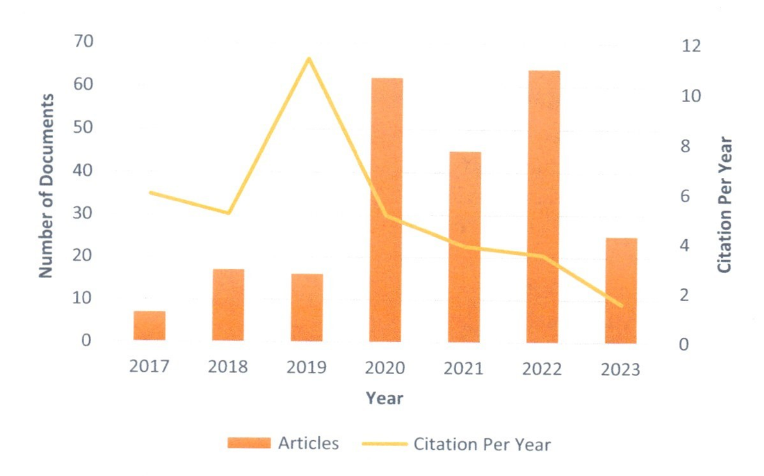

Fig. 10.2 Annual production and citation of documents over the period

Table 10.2 Most relevant and cited journals

Journal name	h_index	TC	NP	PY_start
Technological Forecasting and Social Change	11	505	14	2020
Journal of Industrial Integration and Management-Innovation and Entrepreneurship	8	377	10	2018
Asia Pacific Journal of Innovation and Entrepreneurship	5	194	6	2017
Business Horizons	5	523	5	2017
IEEE Access	4	76	7	2020
IEEE Transactions on Engineering Management	4	120	5	2020
Journal of Entrepreneurship and Public Policy	4	42	8	2020
Small Business Economics	4	141	6	2017
Entrepreneurship and Sustainability Issues	3	19	7	2019
Journal of Enterprise Information Management	3	18	3	2020
Research Policy	3	55	3	2020
2019 42nd International Convention on Information and Communication Technology, Electronics and Microelectronics (MIPRO)	2	17	2	2019
2020 43rd International Convention on Information, Communication and Electronic Technology (MIPRO 2020)	2	6	8	2020
Electronic Commerce Research and Applications	2	64	2	2020
European Journal of Finance	2	81	2	2021
International Journal of Entrepreneurship and Innovation	2	11	2	2019
International Journal of Production Research	2	17	2	2021
International Journal of Technology Management	2	8	2	2020
Journal of Business Research	2	69	3	2021
Journal of Industrial and Business Economics	2	35	2	2019

important and relevant sources for the research study. As shown in Table 10.2, the *Technological Forecasting and Social Change*, with 505 citations, 14 publications, and its first publication in 2020, has the highest h-index (11), followed by the *Journal of Industrial Integration and Management-Innovation and Entrepreneurship* (h-index 8), with 377 citations, 10 publications, and its first publication in 2018. Other prominent journals include the *Asia Pacific Journal of Innovation and Entrepreneurship* (194 citations, 6 publications, and its first publication in 2017) and *Business Horizons* (523 citations, 5 publications, and its first publication in 2017), with h-index 5 and similar ranking.

3.1.4 Most Relevant Authors

The top 20 authors in the research under study are listed together with their TC, NP, and PY-start scores (Table 10.3). Figure 10.3 depicts their production over time. In Fig. 10.3, the spheres' colour intensity is connected to TC annually, whereas their volume is related to NP annually. According to Table 10.3, the top three ranking

Table 10.3 Most relevant authors

Element	h_index	TC	NP	PY_start
Islam N	5	113	5	2020
Mondal S	3	47	3	2021
Novak M	3	61	3	2020
Paul T	3	47	3	2021
Potts J	3	50	4	2020
Rakshit S	3	47	3	2021
Allen DWE	2	47	3	2020
Berg C	2	46	4	2020
Chen Y	2	333	3	2018
Dong J	2	13	2	2021
Fisch C	2	28	3	2022
Gorkhali A	2	23	2	2022
Grobys K	2	12	2	2021
Gunasekaran A	2	35	2	2022
Huang GQ	2	23	2	2020
Johan S	2	34	2	2019
Kumar S	2	14	2	2022
Lu Y	2	103	2	2018
Marinakis Y	2	66	2	2020
Markey-Towler B	2	49	3	2020

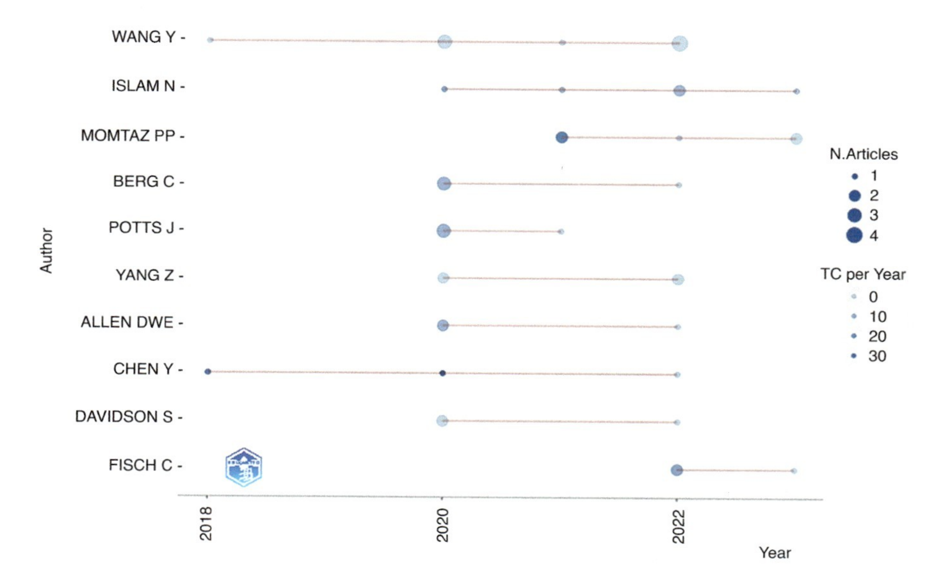

Fig. 10.3 Production over time of relevant authors

authors by h-index are ISLAM N (with 5 publications, an h-index of 5, and 113 citations), MONDAL S (with 3 publications, an h-index of 3, and 47 citations, and their first publication in respective research in 2021), NOVAK M (with 3 publications, an h-index of 3 and 61 citations, and their first publication in blockchain and entrepreneurship research in 2020).

3.1.5 Most Contributed Countries

There have been 60 nations in total that have published journal publications on this subject. According to Single Country Publications (SCP) and Multiple Country Publications (MCP), Fig. 10.4 shows which countries produce the most in the fields of blockchain technology and entrepreneurship research. The top two countries in terms of SCP are China and the USA, with Australia, India, Germany, and Croatia rounding out the top five, whereas in terms of MCP, the top two countries are the USA and the United Kingdom, followed by China, India, and Greece. Table 10.4 illustrates the top 10 nations for research on blockchain technology and entrepreneurship based on the number of publications (f) obtained from a study of the country's scientific output. With 172 and 148 publications each, respectively, China and the United States are the top two producers of scientific literature. The UK is next with 68 publications, followed by India with 65 and Germany with 47.

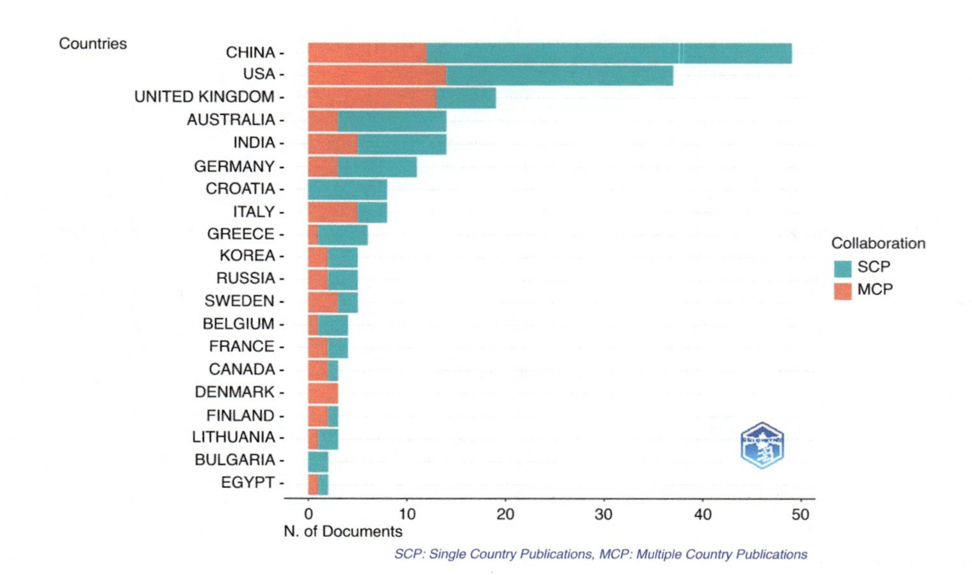

Fig. 10.4 Contribution of countries in terms of publications

Table 10.4 Top 10 contributed countries

Region	Frequency
China	172
USA	148
UK	68
India	65
Germany	47
Australia	34
Italy	27
Croatia	21
France	20
Greece	20

3.2 *Content Analysis*

3.2.1 Keyword Co-occurrence Network Analysis

According to Bagagelj and Cerinek [17], a keyword co-occurrence analysis carried out in R (shown in Fig. 10.6) provides a co-occurrence network representation of the keyword universe and a much better understanding of the dynamics of interaction in the field of blockchain technology and entrepreneurship. A pair of words occurring together is represented as a link in a keyword co-occurrence network, where each keyword is represented as a node. The frequency with which a word pair appears in various publications determines the strength of a link between two terms [18].

The six clusters were identified through keyword co-occurrence analysis depicted in Fig. 10.5. The first cluster (purple) represents the blockchain and its area when it came to existence, that is, bitcoin, cryptocurrency, and smart contracts. The terms in the second cluster (navy) represent the role of blockchain in various fields such as entrepreneurship, crowdfunding, entrepreneurial finance, fintech, governance, etc. The terms related to evolution and implementation covered in the third cluster (green) are strategy, innovation, business model, platforms, IoT, industry, etc. The fourth cluster (red) represents the area and features of blockchain technology, that is, artificial intelligence, supply chain, sustainability, industry 4.0, logistics, big data, security, networks, etc. Clusters five (yellow) and six (blue) represent terms such as trust, design, integration, knowledge, management, performance, etc.

3.2.2 Thematic Map

One of the main benefits of the thematic map is the capability to identify the intensity of research in various groups based on the levels of prominence and density [19]. The density of a theme reveals its evolution, while a theme's centrality is determined by how closely related various themes are to it [20].

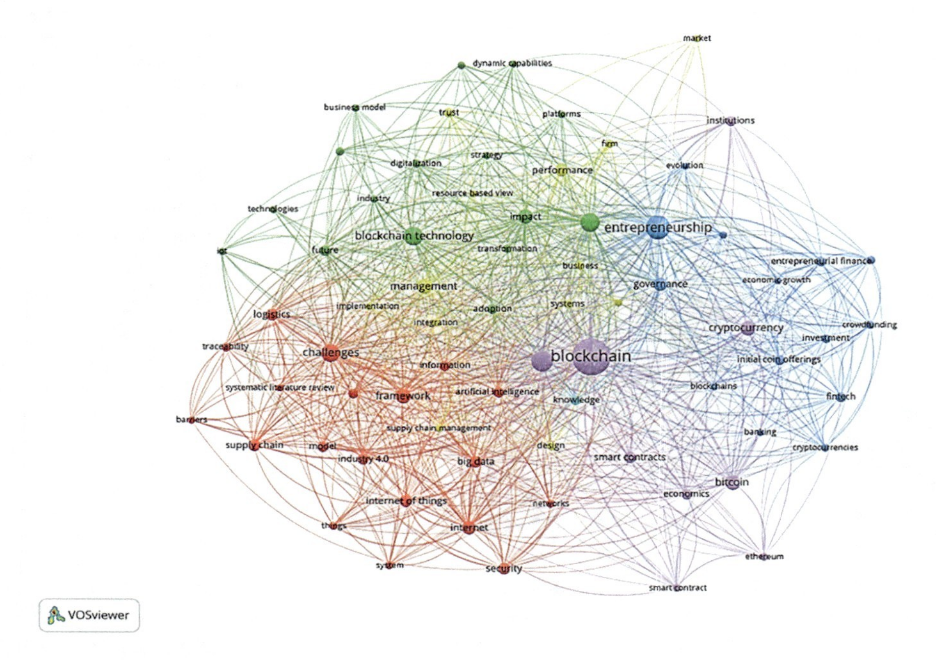

Fig. 10.5 Keyword co-occurrence network

A thematic map (Fig. 10.6) depicts that blockchain technology and cryptocurrencies are the niche theme also, theme with which the technology came into existence in the beginning. However, its impact and adoption theme fall in declining themes, as the adoption of technology has been made and explored in many domains, but the impact and utilization are yet to be explored. The challenges, framework, and management are the next step after the adoption of any technology, and these themes fall under basics, whereas entrepreneurship and innovation along with technology are motor themes.

3.2.3 Authors Co-citation Network

How recurrently two documents are cited alongside one another is revealed by co-citation analysis. When these documents are mentioned together more frequently, their strength grows [21]. Through an author co-citation network, researchers can gradually discover the most well-known experts in a certain field of knowledge [22]. The co-citations between the authors were explored and visualized using VOSviewer. The selection of authors and the co-citation analysis first took place in VOSviewer.

A total of 10,055 writers were listed; however, only 37 of them met the requirement of having 15 or more citations. In the co-citation network, they were represented by 37 nodes grouped into four clusters (Fig. 10.7). The most three prominent authors based on their citations have been identified from each cluster. Cluster

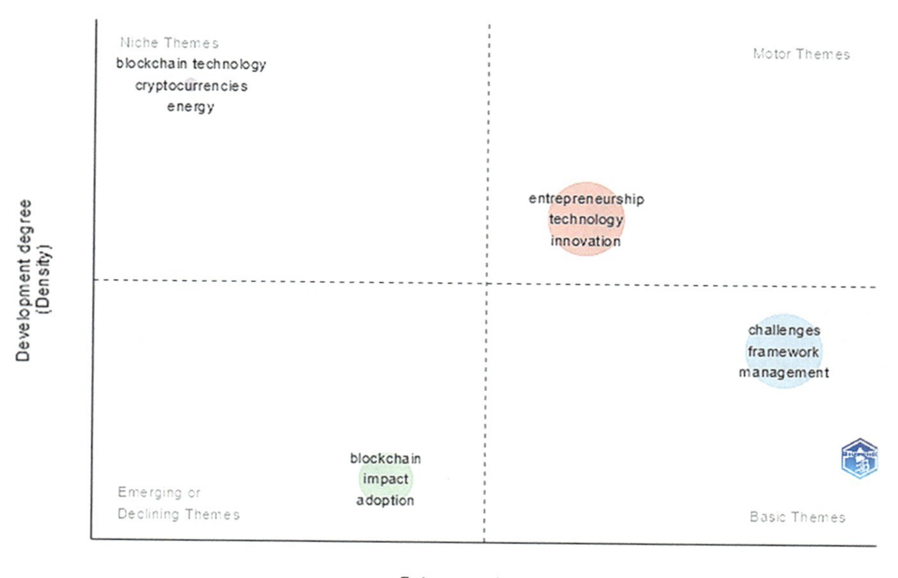

Fig. 10.6 Thematic map of blockchain technology and entrepreneurship research

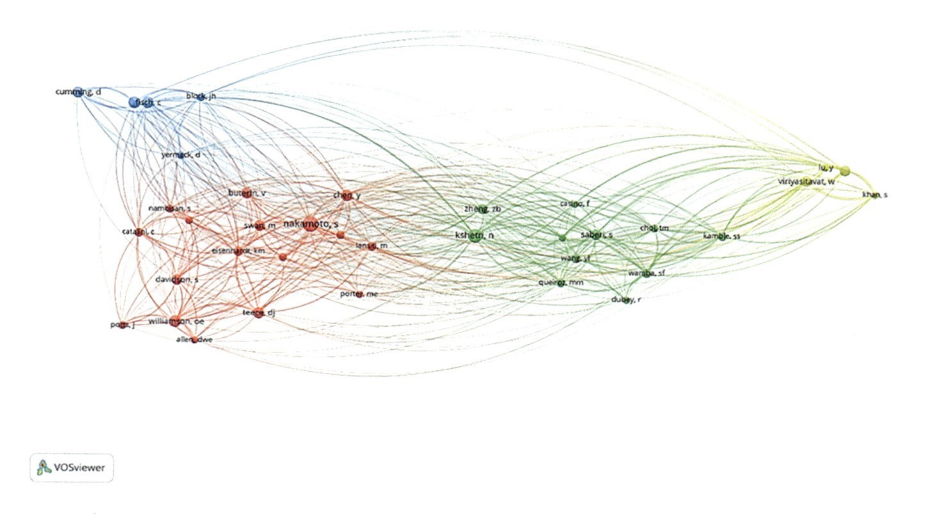

Fig. 10.7 Author's co-citation network

one(red): Nakamoto, S (61 citations and 264 link strength), Williamson, O.E (35 citations and 239 link strength), and Teece, D.J (34 citations and 164 link strength). The second cluster (green): Kshetri, N (48 citations and 302 link strength), Zheng, Z.B (25 citations and 130 link strength), and Saberi, S (28 citations and 206 link strength). The third cluster (blue): Fisch, C (36 citations and 276 link strength), Momtaz, P.P (36 citations and 258 link strength) and Cumming, D (32 citations and 128 link strength). Fourth cluster (yellow): Viriyasitavat, W (31 citations and 364 link strength), Xu, I.D (29 citations and 261 link strength), and Lu, Y (18 citations and 178 link strength).

3.2.4 Countries Collaboration Worldmap

On a worldmap, Fig. 10.8 depicts global trends in research collaboration. The most productive partnership is between the United States and the United Kingdom ($f = 13$), followed by China and the United States ($f = 10$), and the United States with India and Germany ($f = 6$). Table 10.5 shows the top ten collaborations and the number of works produced. This evidence implies that geographical or linguistic proximity does not affect global collaborative networks.

4 Discussion

The study examined blockchain technology in relation to entrepreneurship in the subject area of Business, Management, Economics, Business Finance, Operations Research Management, and Telecommunications using bibliometric analysis. The

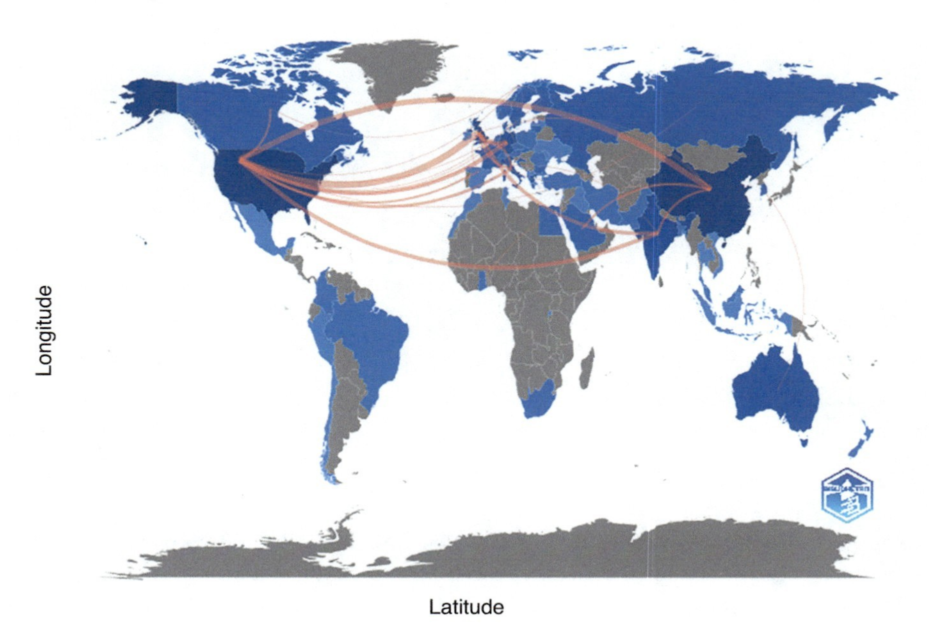

Fig. 10.8 Worldmap of countries' collaboration

Table 10.5 Top 10 collaborations of countries

From	To	Frequency
USA	United Kingdom	13
China	USA	10
USA	India	6
USA	Germany	6
United Kingdom	India	5
United Kingdom	Germany	5
China	India	3
United Kingdom	Italy	3
United Kingdom	Belgium	3
USA	Canada	3

analysis showed that there is no chronological upward and downward trend of publication in the field. However, 2022 is having 64 publications followed by 2020 with 62 publications. The first research on blockchain technology regarding entrepreneurship was made in 2017 even though the concept of blockchain was introduced in 2008. The de Leon discussed the properties of blockchain and the misconceptions about them. Even though the most three prominent authors based on h-index are ISLAM N, MONDAL S, and NOVAK M. Islam N studied whether the blockchain is advantageous in the long run and discovered that, despite initial profitability, blockchain miners cannot maintain long-term financial viability without significant fees. He also examined in one of the studies the impact of blockchain technology on SMEs internationalization [23]. Mondal S and Islam N [24] found in their research that blockchain technology has a positive impact on the supply chain of tea and transparency and reliability are the most important factors of sustainable performance, whereas Novak [25] discussed about the blockchain public policy. Based on the h-index, the most three prominent journals are *Technological Forecasting and Social Change*, *Journal of Industrial Integration and Management-Innovation and Entrepreneurship*, and *Journal of Industrial Integration and Management-Innovation*. The whole world is contributing to the research, but concerning blockchain technology in entrepreneurship, China, the USA, and Australia are the most productive countries in terms of SCP, and in terms of MCP, the USA, United Kingdom, and China are the ones.

The thematic analysis identified entrepreneurship, technology, and innovation based on the frequency of occurrence as motor themes. Blockchain is one of the most emerging applications with a significant potential for disruption, yet many parts of this invention remain unknown in the present corpus of the research [26].

Furthermore, according to the co-citation analysis of authors, Nakamoto S is highly cited by other authors. However, no significant and direct research has been performed by the author. The author proposed the peer-to-peer network for electronic cash transfer [27], which was used as the foundation for so many studies in blockchain technology.

Finally, the countries collaboration worldmap indicates that the United States is the country that provides the most blockchain technology in entrepreneurship research, as well as strong collaborations with other countries such as the United Kingdom, India, and Germany.

4.1 Theoretical Implication

Blockchain technology in entrepreneurship is an emerging topic that has captivated the interest of a diverse group of people, including researchers, academics, and research institutes. From a scholarly viewpoint, the bibliometric study that was conducted provides some results concerning crucial issues that academics should take into account while blockchain technology in entrepreneurship. Academics can understand the context of the many factors that have contributed to the interdisciplinary study of blockchain technology. This study can help researchers identify and comprehend new themes, as well as the keywords employed, literature that addressed these topics, and relevant references. It offers insight into the topic's importance and can therefore be used as a starting point for additional research and to understand various factors such as innovation, transparency, sustainability in context with blockchain technology in entrepreneurship.

4.2 Managerial Implications

The concept of blockchain plays a very important role in increasing the efficiency and transparency in entrepreneurship and also in its sustainable performance. By taking this study as a base, professionals can explore the various factors affecting entrepreneurial performance by using blockchain technology.

5 Conclusion

The study concluded by conducting an extensive bibliometric examination of the application of blockchain technology in entrepreneurship across numerous areas. The results of the analysis showed that there is no discernible chronological trend in the appearance of research articles in this field. However, 2022 and 2020 were closely followed by 2022 in terms of publications. It is interesting to note that despite the blockchain concept being presented in 2008, the first research on the topic of entrepreneurship and blockchain technology has been conducted in 2017.

Based on their h-index, prominent authors like ISLAM N, MONDAL S, and NOVAK M appeared, demonstrating their considerable contributions to the subject. The survey also highlighted the top countries and publications that contributed to blockchain technology in entrepreneurship research. The three journals with the highest h-indexes were *Technological Forecasting and Social Change*, *Journal of Industrial Integration and Management-Innovation and Entrepreneurship*, and *Journal of Industrial Integration and Management-Innovation*. In terms of research production and collaboration, the United States served as a catalyst among the nations, closely followed by the United Kingdom, China, and Australia.

Overall, the study highlighted the novelty of blockchain technology in entrepreneurship research, highlighting its potential for disruption as well as the need for more research and understanding in this rapidly evolving area.

5.1 Limitation and Future Research

One of the paper's limitations, despite its contributions, is the use of a single database rather than numerous sources to retrieve data. Also, the data were limited to a few subject areas to focus only on entrepreneurship, and the future researchers can explore other databases and other domains. Only VOSviewer and the R program were used to analyze this study. Other tools, such as Bibexcel, Gephi, Tableau, and CiteSpace II, can be utilized in future studies in addition to the R package and VOSviewer. Future research may also do factorial analysis, co-citation analysis of sources and documents, historiography, author collaboration network, and other descriptive analyses to further explore the subject.

References

1. Yli-Huumo, J., Ko, D., Choi, S., Park, S., & Smolander, K. (2016). Where is current research on blockchain technology? – A systematic review. *PLoS One, 11*(10), e0163477.
2. Reyna, A., Martín, C., Chen, J., Soler, E., & Díaz, M. (2018). On blockchain and its integration with IoT challenges and opportunities. *Future Generation Computer Systems, 88*, 173–190.
3. Andoni, M., Robu, V., Flynn, D., Abram, S., Geach, D., Jenkins, D., & Peacock, A. (2019). Blockchain technology in the energy sector: A systematic review of challenges and opportunities. *Renewable and Sustainable Energy Reviews, 100*, 143–174.
4. Feng, Y., Zhong, Z., Sun, X., Wang, L., Lu, Y., & Zhu, Y. (2023). Blockchain enabled zero trust-based authentication scheme for railway communication networks. *Journal of Cloud Computing, 12*(1), 1–21.
5. Idrees, S., & Nowostawski, M. (2022). *Transformations through blockchain technology*. Springer.
6. Abdollahi, A., Sadeghvaziri, F., & Rejeb, A. (2023). Exploring the role of blockchain technology in value creation: A multiple case study approach. *Quality and Quantity, 57*(1), 427–451.
7. Insights, D., Pawczuk, L., Holdowsky, J., Massey, R., & Hansen, B. (2020). *Deloitte's 2020 global blockchain survey: From promise to reality*.
8. Rawhouser, H., Vismara, S., & Kshetri, N. (2023). Blockchain and vulnerable entrepreneurial ecosystems. *Entrepreneurship and Regional Development*, 1–26.
9. Acs, Z. J., Song, A. K., Szerb, L., Audretsch, D. B., & Komlosi, E. (2021). The evolution of the global digital platform economy: 1971–2021. *Small Business Economics, 57*, 1629–1659.

10. Autio, E., Nambisan, S., Thomas, L. D., & Wright, M. (2018). Digital affordances, spatial affordances, and the genesis of entrepreneurial ecosystems. *Strategic Entrepreneurship Journal, 12*(1), 72–95.
11. Donthu, N., Kumar, S., Mukherjee, D., Pandey, N., & Lim, W. M. (2021). How to conduct a bibliometric analysis: An overview and guidelines. *Journal of Business Research, 133*, 285–296.
12. Ardito, L., Scuotto, V., Del Giudice, M., & Petruzzelli, A. M. (2019). A bibliometric analysis of research on Big Data analytics for business and management. *Management Decision, 57*(8), 1993–2009.
13. Riahi, Y., Saikouk, T., Gunasekaran, A., & Badraoui, I. (2021). Artificial intelligence applications in supply chain: A descriptive bibliometric analysis and future research directions. *Expert Systems with Applications, 173*, 114702.
14. Kim, H., & So, K. K. F. (2022). Two decades of customer experience research in hospitality and tourism: A bibliometric analysis and thematic content analysis. *International Journal of Hospitality Management, 100*, 103082.
15. van Raan, T. (2014). Advances in bibliometric analysis: Research performance assessment and science mapping. In W. Blockmans (Ed.), *Bibliometrics: Use and abuse in the review of research performance* (Vol. 87). L Engwall and D Weaire Portland Press Ltd.
16. Yang, K., & Meho, L. I. (2006). Citation analysis: A comparison of Google scholar, scopus, and web of science. *Proceedings of the American Society for Information Science and Technology, 43*(1), 1–15.
17. Batagelj, V., & Cerinšek, M. (2013). On bibliographic networks. *Scientometrics, 96*, 845–864. https://doi.org/10.1007/s11192-012-0940-1
18. Radhakrishnan, S., Erbis, S., Isaacs, J. A., & Kamarthi, S. (2017). Novel keyword co-occurrence network-based methods to foster systematic reviews of scientific literature. *PLoS One, 12*(3). https://doi.org/10.1371/journalpone0172778
19. Mobin, M. A., Masnun Mahi, M., Hassan, K., Habib, M., Akter, S., & Hassan, T. (2021). An analysis of COVID-19 and WHO global research roadmap: Knowledge mapping and future research agenda. *Eurasian Economic Review*. https://doi.org/10.1007/s40822-021-00193-2
20. Esfahani, H. J., Tavasoli, K., & Jabbarzadeh, A. (2019). Big data and social media: A scientometrics analysis. *International Journal of Data and Network Science, 3*(3), 145–164. https://doi.org/10.5267/ijdns20192007
21. Pelit, E., & Katircioglu, E. (2022). Human resource management studies in hospitality and tourism domain: A bibliometric analysis. *International Journal of Contemporary Hospitality Management, 34*(3), 1106–1134. https://doi.org/10.1108/IJCHM-06-2021-0722
22. Lin, J.-S., & Himelboim, I. (2018). Political brand communities as social network. *Journal of Political Marketing*. https://doi.org/10.1080/15377857201814786661
23. Rakshit, S., Islam, N., Mondal, S., & Paul, T. (2022). Influence of blockchain technology in SME internationalization: Evidence from high-tech SMEs in India. *Technovation, 115*, 102518.
24. Paul, T., Mondal, S., Islam, N., & Rakshit, S. (2021). The impact of blockchain technology on the tea supply chain and its sustainable performance. *Technological Forecasting and Social Change, 173*, 121163.
25. Novak, M. (2020). Crypto-friendliness: Understanding blockchain public policy. *Journal of Entrepreneurship and Public Policy, 9*(2), 165–184.
26. Laaraj, M., Nakara, W. A., & Fosso Wamba, S. (2022). Blockchain diffusion: The role of consulting firms. *Production Planning and Control*, 1–13.
27. Nakamoto, S. (2008). Bitcoin: A peer-to-peer electronic cash system. *Decentralized Business Review, 21260.*

Chapter 11
Transforming Educational Landscape with Blockchain Technology: Applications and Challenges

Roshan Jameel, Bhawna Wadhwa, Alisha Sikri, Sachin Singh, and Sheikh Mohammad Idrees

1 Introduction

The blockchain technology was introduced in 2008 as a distributed ledger for maintaining the cryptocurrency called Bitcoin [1]. The goal of proposing such a network was to eliminate the need of third party for the verification of transactions and make the communication channel direct and transparent. The blockchain network is designed as a collection of connected nodes each of which has a copy of the ledger. Whenever a node wants to write something on the network or carry out the transaction, the transaction is broadcasted to all the nodes within the network, and the transaction is only considered as complete if the consensus is achieved. Furthermore, the nodes are also responsible for regularly checking the status of the ledger and its own copy to make sure that the data integrity is maintained all the time [2]. The immutability and transparency of Blockchain technology makes it a trustworthy network for several potential application domains [3, 23]. The blockchain technology became popular because of its ability to handle sensitive data and to revolutionize various application areas including healthcare, finance, education, Internet of Things (IoT), etc. at global level [4, 5, 24].

Currently, the blockchain technology has been applied in cryptocurrencies like Bitcoin, Ethereum, Zcash, etc. Bitcoin is the first ever peer-to-peer e-cash network that maintains consensus for carrying out transactions. Whereas, Ethereum and Zcash are open-source platforms for hosting cryptocurrencies [6]. Several researchers have divided the development of blockchain technology into three stages: Blockchain 1.0 involved the development and deployment of cryptocurrencies;

R. Jameel (✉) · B. Wadhwa · A. Sikri · S. Singh
Noida Institute of Engineering and Technology, Noida, India

S. M. Idrees
Norwegian University of Science and Technology, Gjøvik, Norway

S. M. Idrees, M. Nowostawski (eds.), *Blockchain Transformations*, Signals and Communication Technology, https://doi.org/10.1007/978-3-031-49593-9_11

Blockchain 2.0 introduced the concepts related to smart contracts, smart properties, bonds etc.; while Blockchain 3.0 is in the developing phase of the technology that applies to various application domains beyond finance and banking [7, 8]. There are several domains of application of the blockchain technology, including healthcare, government, logistics, identity management, higher education, etc. [9]. By integrating blockchain in the education sector, the way of storing, sharing, and verifying the academic credentials can be transformed. With the help of blockchain, the institutes can develop a system that is tamperproof for keeping the student records and certificates, verifying credentials, automating administrative tasks, etc.

2 Blockchain Technology

Blockchain is a technology based on a distributed ledger that makes the transactions and transmission of assets among the users secure and transparent at a low cost [10]. The data flow of a blockchain transaction is shown in Fig. 11.1 below as follows: When a node wants to initiate a transaction through the blockchain ledger, its identity is verified first. The transaction is then broadcasted to the entire network and waits for it to get validated from the nodes. After the transaction is verified and validated, a block is generated and all the nodes on the network get an updated copy of the blockchain. Such blocks are generated after every unit of time and are added to the blockchain through cryptographic digital signatures [11]. The validation of the

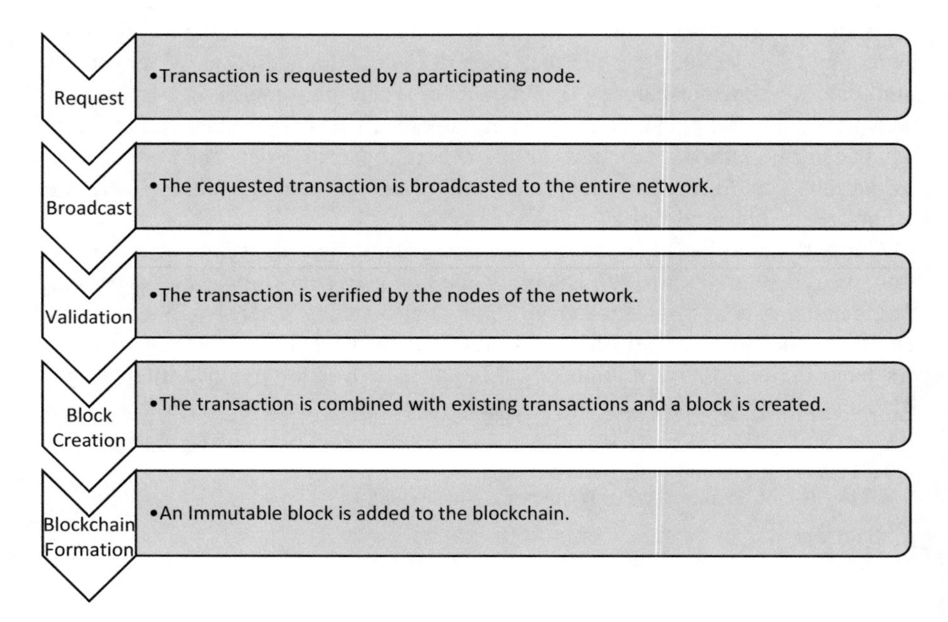

Fig. 11.1 Data flow in a blockchain network

transactions is done via consensus mechanisms called mining. The mining is done using the computational capacities of the CPUs and the nodes decide whether a new block can be added to the blockchain or not. Each of the transactions within the block and the linking of the two blocks are both timestamped. Therefore, the data on the blockchain is aligned in terms of time and the blockchain is always growing with time. The data on the blockchain is immutable, that is any data once stored on the chain can never be updated or deleted. However, in any circumstance, if it is required to be updated, a new transaction is generated and added to the blockchain, and the older one is also kept intact.

3 Properties of Blockchain Technology

There are several prominent properties of blockchain technology that make it unique and apt for various application areas. Some of them are shown in Fig. 11.2 and discussed below:

Decentralization Blockchain technology supports decentralization, in which there is no central authority like government or third party for validating or controlling the network. The geographically distributed computers connected to one another are

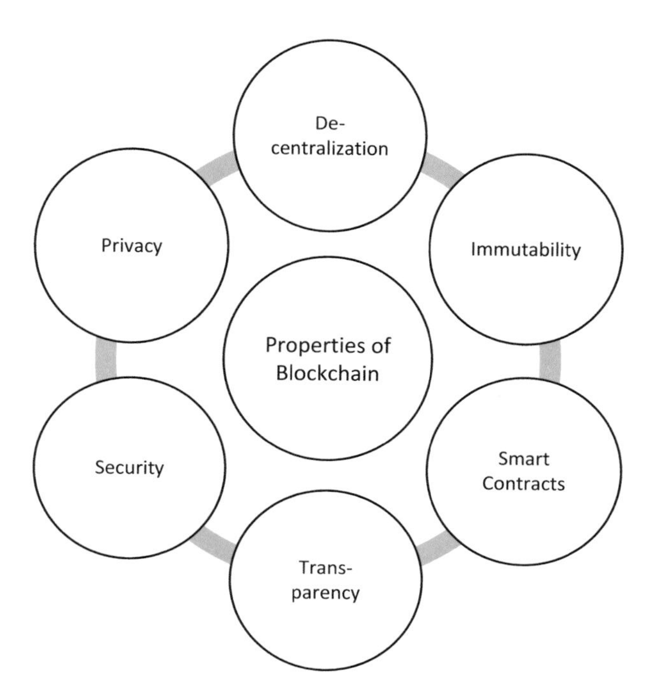

Fig. 11.2 Properties of blockchain technology

called nodes. The validation of the transactions is done through consensus algorithms.

Immutability When the data is added on the blockchain, it becomes permanent, i.e., it's not possible to modify that data in any manner. The blocks contain the hash value of the content of that block and the hash value of the previous block, forming a link between the two blocks. This connection makes it impossible to alter the data on the block, thus assuring the data integrity and immutability.

Smart Contracts These are the automatically executable set of rules that are aligned with the blockchain network. These contracts are self-executed and enforced and eliminate the need of intermediaries, thus reducing the chances of fraud and developing trust among the participating parties.

Transparency In blockchain, transparency is provided by the network by letting all the nodes of the network have access to the same information. The transactions taking place on the network are made visible to all the nodes making the entire process auditable and transparent.

Security Advance cryptographic algorithms and digital signatures are implemented in blockchain to provide security. Furthermore, the decentralized alignment of the nodes makes the network resistant to fraudulent and unauthorized activities. Also, the consensus mechanisms assure validity of the transactions and enhance the security of the network.

Privacy The blockchain provides privacy to the users by allowing them to participate in the network with pseudonymous identity, and the actual real-world IDs are not directly mapped. Privacy and integrity of the data is assured using cryptographic algorithms, zero-knowledge proofs, etc.

4 Blockchain Technology in Education

Currently, the blockchain concept has been applied by a few educational institutions mainly for managing the degrees and certificates or for evaluating the outcomes of their courses [12, 13]. However the blockchain in the education sector can be applied in formal as well as informal settings. The formal framework includes the content of teaching, outcomes of students, certificates and degrees, etc. Whereas, the informal framework consists of research and skillset-related information. The blockchain-based network can help in matching all types of information about the user with his ID, such as academic certifications, project-related information, behavioral traits, etc. There are several use cases of blockchain technology in the education sector; some of the prominent ones are shown in Fig. 11.3 and discussed below:

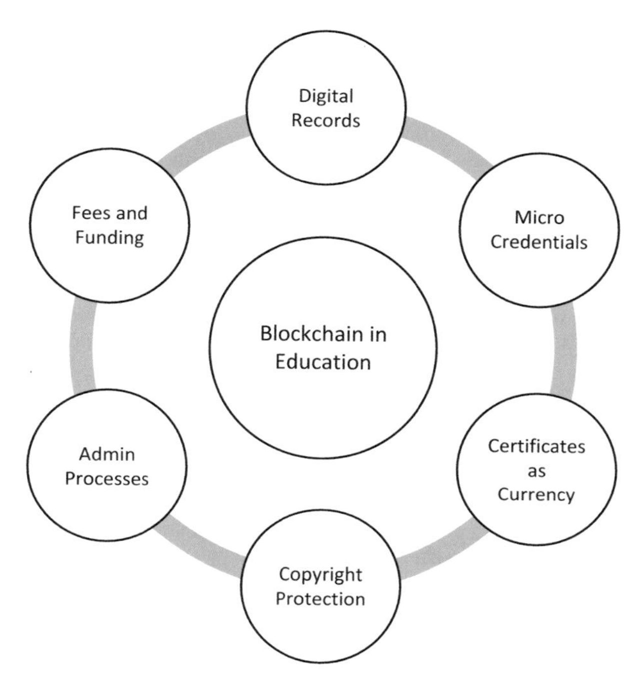

Fig. 11.3 Use cases of blockchain technology in education

Digital Records and Transcripts The blockchain can be used to store transcripts, degrees, diplomas, certificates of the students in digital form, so that it can be accessed by the students, academicians, or any other authorized third party. Having the records stored on the blockchain would provide the students with the availability of the data that is immutable without the need of any administrative intermediary. The blockchain is a decentralized ledger, that keeps the data available all the time as a copy is maintained by all the nodes of the network. In case of failure or unavailability of a node, the data can be accessed all the time. The blockchain-based network for storing such data would also save the processing and verification time of issuing the documents. Furthermore, the data on blockchain is immutable, which means the certificates and degrees will remain unharmed and unmodified during their lifetime. The cases of degree fraud can also be eliminated as the data stored on the blockchain will be verified and validated by the miners throughout the globe.

Micro Credentialing The blockchain can be facilitated to validate the micro-credentials and badges that are earned by the students through small courses or online certifications etc. These credentials can be stored as proof on the blockchain in a transparent and immutable manner. The University of Nicosia is the first ever educational firm that implemented blockchain network for managing their MOOC-related content [12]. Another organization is Sony Global Education, which uses the technology to store and manage the degrees [14]. Another such concept was applied by the Massachusetts Institute of Technology (MIT) along with a machine learning–based company to provide a digital badge to the students who are attending some online certification courses [13].

Certifications as Currency Another way of leveraging the blockchain technology in the educational domain is to use it as a capacity currency. In this concept, the learning experience of the users, their knowledge and skillset, etc. can be transformed into digital tokens or currencies that can be stored on a wallet as per the standards. The students will get rewarded on the basis of their efforts and accomplishments in the form of these tokens only which will be added to their wallets [12]. This concept has been applied with the currency name 'Kudos', which is stored on a digital and measures the learning hours of the users.

Copyright Protection The blockchain technology can be utilized as a protector of the intellectual properties and copyrights of individuals. For instance, it can be used to keep the academic details and research-related information in a timestamped and transparent manner. Thus, providing the evidence to support the authorship and ownership of the research, it will also keep track of the research-related activities in a timely manner. Furthermore, plagiarism can also be controlled by providing a secure way of sharing, accessing and transmitting the information among the different users while maintaining the control of the owners on their data.

Administrative Processes By implementing the blockchain-based distributed ledger in the educational sector, several administrative tasks can be automated in a transparent, secure and errorless manner. Tasks such as student registration, enrollment, payment verification, degree issuance, etc. can done by applying smart contracts, thus reducing the intermediaries, paperwork and human errors, which will save time and effort of both students and the institutions. The blockchain-based network will keep track of availability of courses or seats in a particular course, keep the documents and certificates of the students in a digital wallet, verify academic credentials, and implement supply chain management of educational resources, royalty and copyright management, etc.

Fees and Funds Blockchain has the potential to transform the entire process of managing the fees and funding of educational institutions. The blockchain-based ledger for keeping the records is immutable, which allows the institutions to keep track of the transactions and also helps students in verifying the payments. The blockchain helps in the process of application and distribution of scholarships and eliminates unfair activities. Likewise, funding for researches and projects can also be traced for their genuine usage. Integrating the blockchain in fees and funding management can improve accountability, efficiency and transparency among the different stakeholders.

Table 11.1 below summarizes the features of Blockchain technology and how it can be leveraged in various areas within the education sector:

Table 11.1 Blockchain features and implementation in education sector

Feature	Description	Implementation in education sector
Decentralization	The blockchain-based ledger is distributed, i.e., there is no central authority that controls the network	Implementing the decentralized blockchain network will allow the verification of the academic badges, identities, transcripts, etc., thus reducing the requirement of intermediaries. It ensures the ownership of the data and distribution among different stakeholders
Immutability	The data once stored on the blockchain can neither be altered nor be deleted	The blockchain network is immutable; records once created cannot be deleted or changed. This ensures the integrity of the data along with insurance against fraudulent claims related to fees, scholarship and funding. It also verifies the certificates or badges earned by a student thus, preventing fraud degree scams.
Smart contracts	Blockchain-based applications work on automated set of rules and regulations	Smart contracts can be used to automate the administrative processes in the educational institutions. This will reduce time consumption, human error and paperwork
Transparency	The transactions on the blockchain network are visible to all the nodes on the network	The student records, certifications, badges and achievements can be stored in the blockchain network, and their validity can be ensured by blockchain verification by employers, institutions, etc., hence enhancing the trust of the stakeholders
Security	Cryptographic algorithms are applied in blockchain, preventing them from unauthorized access and making them difficult to hack	Sensitive information such as personal details of the student and academic records are secure against data and identity thefts in a blockchain-based network
Privacy	The blockchain supports the concept of pseudonymous identities, which keeps the user's actual identities private	The blockchain offers pseudo identities to its users, which are not related to their real-world identities at all. This feature ensures the privacy of the users alongside the integrity of their data

5 Challenges in Integrating Blockchain and Education

The blockchain has become popular in the last decade, mostly it is portrayed as a magical technology that is the solution of every application domain because of the transparency and security it provides. However, the technology has its own challenges, such as scalability issues, loss of control over data, power and resources consumption, lack of required skills and expertise, to name a few. In order to adopt this emerging technology, one must keep in mind these challenges and analyze the feasibility before expecting the unrealistic services. This section discusses and Fig. 11.4 depicts some of the challenges faced while integrating the blockchain technology in the education sector.

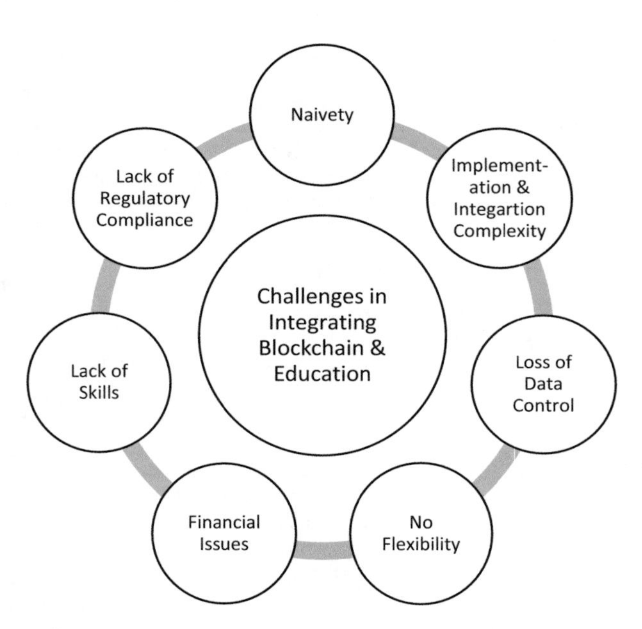

Fig. 11.4 Challenges of integrating blockchain technology in education

Naivety The blockchain technology is still evolving and immature. Several articles have been published that discuss the sufferings of blockchain-based applications because of its naivety [15–17]. Since the blockchain is still in its developing stage it goes through several problems, such as poor usability, lack of standardization, lack of interoperability, scalability issues, etc. The naivety in the blockchain can be seen from two perspectives, the users and the developers. The developers are not very skilled whereas, the end users are expecting that moving any application to the blockchain would be sufficient to solve all the associated problems. Therefore, the users as well as the organizations are supposed to thoroughly do the feasibility study before moving their business on the blockchain technology.

Implementation and Integration Complexity In order to move the educational systems completely on the blockchain-based network, the institutions have to move their existing infrastructure and reorganize everything entirely onto a new network. The educational infrastructures are usually very complex and consist of various components, such as software, databases and data types. Several articles have discussed problems related to the aligning, integrating and interfacing between the systems. Therefore, the complexities associated with implementation and integration of blockchain must be kept in mind before adopting. A well-planned approach should be followed for seamlessly integrating the blockchain to the existing infrastructure.

Loss of Data Control The blockchain technology provides security to the data as it is distributed among the nodes of the network. However, while providing data security, the control of the user on its data is lost. And when it comes to educational data such as degrees, marksheets and certificates, issues related to privacy arise [18]. The academic data is sensitive and the transactions on the blockchain are broadcasted to everyone, the institutions as well as the students trust the network with their valuable data, and if this data is kept on the ledger that is publicly available, then no privacy is preserved. Therefore, it is suggested that academic institutions must adopt private blockchain [19]. Furthermore, regulations must also be applied to maintain the privacy of the users. Additionally, the users and their data must be anonymized.

No Flexibility Immutability is one of the prominent features of the blockchain that makes sure that the data on the network must remain unchanged. Most of the properties of the blockchain, such as data confidentiality, security and integrity, are all connected to this property [2]. However, it might become a challenge in academics, because if a data is incorrectly entered on the blockchain, it can never be changed or undone. In fact, a new transaction has to be performed and the transaction with the incorrect data will also remain on the network. This feature undoubtedly increases the integrity of the data and decreases the chances of any kind of fraud on the network, thus making the entire network inflexible.

Financial Issues The adoption and implementation of blockchain technology is costly in terms of transaction fees. When it comes to the education sector, the cost of the transactions should not be overlooked. The consensus algorithms that are said to be the core concept behind developing trust among the different user nodes within the network are the most resource- and energy-consuming processes [20]. Also, the entire data transactions that are happening on the blockchain are maintained on each of the nodes; the systems are required to be of sufficient storage. Furthermore, if we take energy consumption in consideration, with the increasing number of transactions, the energy consumption would also increase [21]. Another aspect related to the finances is the cost of hiring an expert who can implement and work on the blockchain. Since this field is still in its developing stage, experts in this field are still lacking and those who know the technology are charging a lot as compared to normal software developers who are working on traditional technologies.

Lack of Skills The blockchain is still in its developing stage, consequently there is a lack of skilled professionals who have expertise in this domain. This has raised the demand of blockchain developers in almost every sector [20]. Resource persons with a high level of expertise and ability to design and develop a trustworthy environment with all the functionalities and usability are required [21]. Therefore, it can be considered as a challenge in implementing blockchain-based educational systems. Several educational institutes are not adopting the blockchain-based system as they lack skilled and knowledgeable developers who can manage the educational data in an effective manner. One of the ways to integrate the blockchain in the educational system is to provide training on the technology to the existing staff so that things can be maintained on an internal level, without the need of hiring new professionals.

Lack of Regulatory Compliance The lack of regulatory compliances in the blockchain technology is reported to be a challenge while integrating the education system to the blockchain [15]. The blockchain works on the principle of immutability, that is data once entered can never be deleted or changed [15]. However, the General Data Protection Regulation (GDPR) in Europe states that the citizens have the 'right to be forgotten' [15, 22]. The GDPR also supports the controllership over the data, i.e., who as a central controlling authority can have the control over the data, and because of the decentralized nature of the blockchain, this also cannot be handled accountably. These two fundamental rights of the data owners are violated when the data is moved to the blockchain network. Because of the transparency provided by the blockchains, the transactions are visible to every node of the network, which makes it compulsory for the institutions to employ strict protocols and constraints on the level of access to the data. Consequently, in order to integrate blockchain and educational systems, the legal compliances must be taken into account. Table 11.2 below summarizes the challenges discussed above.

The challenges of integrating blockchain technology in the education sector should never be underestimated and must be taken into account. Careful planning and feasibility studies must be performed so that the associated risks can be minimized. Table 11.3 below summarizes the possible solutions to the challenges faced while integrating the blockchain technology in the domain of education.

Table 11.2 Blockchain features and implementation in education sector

Blockchain challenge	Concerns with respect to education sector
Naivety	Unrealistic expectations Overestimation of capabilities Overlooking the vulnerabilities Lack of research
Implementation and integration complexity	Complexity in integrating with the existing systems Integration complexity with various software Complex data sharing Technical expertise required
Loss of data control	Limited control over data Unauthorized data access risk Privacy related concerns Dependency on blockchain platforms
No flexibility	Limited adaptability Unable to update data Inability to customize platforms and protocols Non-flexibility in integrating blockchain with existing processes
Financial issues	Implementation and management costs Cost involved in training and hiring staff Processing fees involved Cost involved in sustaining the blockchain-based systems
Lack of skills	Lack of skills and expertise Insufficient training and upskilling opportunities Difficult to train academicians from non-computer or science background
Lack of regulatory compliances	Difficult to comply with privacy Challenging alignment of blockchain with the existing frameworks Vagueness in regulatory guidelines Potential legal problems associated with the data handling

Table 11.3 Challenges faced in integrating blockchain technology in education sector and proposed solutions

Blockchain challenge	Proposed solutions
Naivety	Education and training: Give in-depth instruction to people and groups so they may better comprehend blockchain technology and its complexities. Proof of concepts (PoCs): Small-scale PoCs can be used as a starting point to obtain practical experience and better understand the technology's potential Collaborate: In the beginning, seek advice and mentorship from seasoned blockchain businesses or specialists
Implementation and integration complexity	Pre-built platforms: To make blockchain deployment and integration less difficult, employ pre-built platforms Consult with experts: Avail the services of blockchain development professionals who have practical knowledge of deploying intricate systems Gradual adoption: To get past the integration difficulties, start with simple blockchain use cases and progressively increase the complexity
Loss of data control	Private blockchain: Use private or consortium blockchains to maintain control over data access and sharing Encryption and access control: While handling sensitive data, use robust encryption methods and granular access controls Hybrid solutions: Use blockchain in such a way that it balances data control and transparency by combining traditional databases and blockchain technology
No flexibility	Smart contracts: To automate procedures and provide flexibility through code updates, use programmable smart contracts Dynamic consensus mechanisms: Employ consensus techniques that enable upgrades and changes without forks. Interoperability: Use cross-chain technologies to facilitate data sharing and communication between various blockchains
Financial issues	Funding sources: Look into several funding possibilities for blockchain projects, such as grants, venture capital, or crowdsourcing Tokenization: Introduce tokens or cryptocurrencies to the network as a way to raise money and streamline transactions Cost analysis: To guarantee that blockchain efforts are financially viable, conduct in-depth cost-benefit studies
Lack of skills	Training programmes: Provide internal or external training courses to upgrade the blockchain technology skills of current personnel To cover the skill gap, recruit seasoned blockchain developers, architects, and experts Collaboration: Form a talent pipeline by collaborating with educational institutions to promote blockchain education
Lack of regulatory compliances	Legal expertise: Work together with legal professionals to manage the difficult regulatory environment of blockchain implementation Compliance frameworks: In order to ensure that regulations are followed, blockchain systems should be designed with compliance frameworks in mind Conduct regular audits: Conduct regular audits to ensure that the blockchain solution is still in compliance with any new legislation that may be implemented

6 Conclusion and Discussion

It can be concluded from the above discussion on the blockchain technology and its applications and integration challenges in education that the blockchain has the ability to transform the several aspects of the education sector, but we need to analyze the challenges and perform thorough feasibility study. By integrating the blockchain technology in the education sector, a network can be delivered that is transparent and decentralized which can enhance the overall security of the academic data along with streamlining the administrative processes. The immutable nature of the blockchain assures the integrity of the academic certificates and records, which reduces the chances of fraudulent degrees or misinterpretation of information. Furthermore, the smart contracts help in automating various aspects, such as fee deposits, course enrolment, reduced human errors and paperwork, etc. Moreover, the anonymization of the users and data is also supported by the blockchain-based systems, which assures privacy.

Despite all the potential use cases and applications of blockchain in the education sector, there are several challenges that might be faced while integrating it in existing educational systems. Since the blockchain is a new technology, it is still in its developing phase because of which it faces issues related to usability, interoperability, scalability, etc. Further, if the educational institutes would move their data or processing on the blockchain network, they would have to reorganize everything. And a well-planned approach should be followed for seamlessly integrating the blockchain to the existing infrastructure. The other challenges include loss of control of data because of the distributed nature of blockchain. The immutability makes it impossible to update or delete the data that violates the right of the citizens to be forgotten. Additionally, lack of compliance and regulatory protocols creates vagueness in guidelines. Other challenges that should not be disregarded are lack of skills and financial barriers. The skilled and expert resources in the field of blockchain are still lacking. Moreover, there is a high cost involved in the implementation and maintenance of the blockchain-based network. In order to integrate it with the educational sector, proper trainings of the academicians are required.

To address the challenges discussed and integrate the blockchain and education domain a collaboration is required among the various stakeholders such as institutions, blockchain experts, policy designers, government, etc. If careful analysis and feasibility study are performed, the integration of blockchain and education would revolutionize the education domain and benefit the students, institutions and the entire educational ecosystem.

References

1. Nakamoto, S. (2008). *Bitcoin: A peer-to-peer electronic cash system*. Available online: https://bitcoin.org/bitcoin.pdf
2. Grech, A., & Camilleri, A. F. (2017). *Blockchain in education*. Publications Office of the European Union.

3. Underwood, S. (2016). Blockchain beyond bitcoin. *Communications of the ACM, 59*(11), 15–17. https://doi.org/10.1145/2994581
4. Salah, D., Ahmed, M. H., & Eldahshan, K. (2020). Blockchain applications in human resources management: Opportunities and challenges. In *Proceedings of the EASE '20: Evaluation and assessment in software engineering, Trondheim, Norway, 15–17 April 2020* (pp. 383–389).
5. Collins, R. (2016). Blockchain: A new architecture for digital content. *EContent, 39*(8), 22–23.
6. Beck, R., Czepluch, J. S., Lollike, N., & Malone, S. (2016). Blockchain – The gateway to trust-free cryptographic transactions. In *Research Papers from ECIS2016*. Istanbul.
7. Gatteschi, V., Lamberti, F., Demartini, C., Pranteda, C., & Santamaría, V. (2018). Blockchain and smart contracts for insurance: Is the technology mature enough? *Future Internet, 10*, 20.
8. Swan, M. (2015). *Blockchain: Blueprint for a new economy* (1st ed.). O'Reilly Media.
9. Devine, P. (2015). *Blockchain learning: can crypto-currency methods be appropriated to enhance online learning?* Presented at the ALT Online Winter Conference 2015, Online, (United Kingdom).
10. Tschorsch, F., & Scheuermann, B. (2016). Bitcoin and beyond: A technical survey on decentralized digital currencies. *IEEE Communication Surveys and Tutorials, 18*(3), 2084–2123. https://doi.org/10.1109/COMST.2016.2535718
11. Yli-Huumo, J., Ko, D., Choi, S., Park, S., & Smolander, K. (2016). Where is current research on Blockchain technology?—A systematic review. *PLoS One, 11*(10), e0163477. https://doi.org/10.1371/journal.pone.0163477
12. Sharples, M., & Domingue, J. (2016). The Blockchain and kudos: A distributed system for educational record, reputation and reward. In *Adaptive and adaptable learning* (pp. 490–496). Springer. https://doi.org/10.1007/978-3-319-45153-4_48
13. Idrees, S., & Nowostawski, M. (2022). *Transformations through Blockchain technology*. Springer.
14. Hoy, M. B. (2017). An introduction to the Blockchain and its implications for libraries and medicine. *Medical Reference Services Quarterly, 36*(3), 273–279. https://doi.org/10.1080/02763869.2017.1332261
15. Steiu, M.-F. (2020). Blockchain in education: Opportunities, applications, and challenges. *First Monday, 25*. https://doi.org/10.5210/fm.v25i9.10654
16. Zheng, Z., Xie, S., Dai, H., Chen, X., & Wang, H. (2017). An overview of blockchain technology: Architecture, consensus, and future trends. In *Proceedings of the 2017 IEEE international congress on big data (BigData congress), Boston, MA, USA, 11–14 December 2017* (pp. 557–564).
17. Williams, P. (2019). Does competency-based education with blockchain signal a new mission for universities? *Journal of Higher Education Policy and Management, 41*, 104–117.
18. Arndt, T., & Guercio, A. (2020). Blockchain-based transcripts for mobile higher-education. *International Journal of Information and Education Technology, 10*, 84–89.
19. Li, H., & Han, D. (2019). EduRSS: A blockchain-based educational records secure storage and sharing scheme. *IEEE Access, 7*, 179273–179289.
20. Delgado-von-Eitzen, C., Anido-Rifón, L., & Fernández-Iglesias, M. J. (2021). Blockchain applications in education: A systematic literature review. *Applied Sciences, 11*, 11811.
21. Haugsbakken, H., & Langseth, I. (2019). The blockchain challenge for higher education institutions. *European Journal of Education, 2*, 41–46.
22. Lin, I. C., & Liao, T. C. (2017). A survey of blockchain security issues and challenges. *International Journal of Network Security, 19*, 653–659.
23. Idrees, S. M., Nowostawski, M., Jameel, R., & Mourya, A. K. (2021). Security aspects of blockchain technology intended for industrial applications. *Electronics, 10*(8), 951.
24. Idrees, S. M., Nowostawski, M., & Jameel, R. (2021). Blockchain-based digital contact tracing apps for COVID-19 pandemic management: Issues, challenges, solutions, and future directions. *JMIR Medical Informatics, 9*(2), e25245. https://doi.org/10.2196/25245. PMID: 33400677; PMCID: PMC7875568.

Chapter 12
Verificate – Transforming Certificate Verification Using Blockchain Technology

Tanmay Thakare, Tanay Phatak, Gautam Wadhani, Teesha Karotra, and R. L. Priya

1 Introduction

Education is an ever-growing industry. Each year, more and more students complete various educational programs and opt for a job or further studies. According to the Ministry of Education, India, approximately 9.54 million students graduated in 2021, which shows an increase of 140,000 students compared to 2020 [1]. Because so many students graduate each year, rigorous screening processes must be put to work to select suitable candidates. As per the World Economic Forum Report published in Nasscom community, the unemployment rate in India during the year 2020 increased to 23% from 10.4% [2]. The consequence of high levels of unemployment is a bunch of candidates trying to delude their eligibility by producing counterfeit certificates. The same applies to higher studies, where candidates fabricate certificates with false results to get admitted to their dream university. According to an annual trends report by AuthBridge, fake degree submissions accounted for about 28% of education disparities in 2020 in India [3].

It becomes a challenge for employers and universities to select valuable candidates from the extensive assortment of applications. They need to perform additional time-consuming procedures to verify the documents of candidates. During our survey to find existing systems that try to solve this issue, we came across some systems like "DigiLocker." DigiLocker is a government of India initiative to promote paperless certificates. It is a centralized system maintained by the government

T. Thakare (✉) · T. Phatak · G. Wadhani · T. Karotra · R. L. Priya
Department of Computer Engineering, Vivekanand Education Society's Institute of Technology, Mumbai, India
e-mail: 2020.tanmay.thakare@ves.ac.in; 2020.tanay.phatak@ves.ac.in; 2020.gautam.wadhwani@ves.ac.in; 2020.teesha.karotra@ves.ac.in; priya.rl@ves.ac.in

S. M. Idrees, M. Nowostawski (eds.), *Blockchain Transformations*, Signals and Communication Technology, https://doi.org/10.1007/978-3-031-49593-9_12

of India. As cyber threats intensify, concerns rise about the data security provided by such centralized systems.

In this chapter, we put forward our proposed system, Verificate, and explain how it attempts to solve the issue of counterfeit certificates in the education sector and job market. The following sections give a complete picture of our project. Section 2 'Background' provides insights into the existing research study carried out related to the topic. Section 3 'Proposed System' describes the solutions to overcome some of the issues in the existing system that we implemented in our system. Section 4 'Methodology' describes how our system functions. Section 5 'Implementation' demonstrates the implementation of Verificate. Section 6 'Conclusion and Future Scope' concludes the chapter with suggestions for further enhancements to our system.

2 Background

Gupta et al. [4] discuss how in-depth knowledge of blockchain technology and its frameworks like Truffle and Ganache helps us address issues related to scalability, cost-effectiveness, and security. Their main aim was to establish an efficient college framework to store student data securely.

Pu et al. [5] aimed to explore the potential of blockchain for digital certificates. They conducted analysis on multiple approaches and mentioned six different case studies for the application of blockchain for digital certificates. The case studies included diamond certificates, COVID-19 certificates, classification society certificates, artwork certificates, educational certificates, and renewable energy certificates. It provides brief information about each case study with very generalized implementation. The research work is mostly focussed on theoretical aspects, leading to beneficial analysis.

Lamkoti et al. [6] focus on generating new certificates with a specific template. The certificates are uploaded to the blockchain by the college itself. However, it does not take into account the old certificates generated by conventional methods. Also, as the templates are predefined, it gives rise to scalability issues as different organizations will have different templates.

Gayathri et al. [7] generated digital certificates and hashed them using a chaotic algorithm. The generated hash is then stored on the local blockchain network. In this research work, verification is done only by comparing the hash values. The work simulates a basic blockchain network, demonstrating its application for certificate storage.

Nyaletey et al. [8] created immutable ledgers and used IPFS to store files. In their system, certificates are uploaded onto the blockchain immediately after being generated by the issuer. Validation of the certificates is done by matching hash values. In this chapter, only those certificates that are directly uploaded by the university are present in the blockchain. This chapter, too, fails to include verification of all the old certificates that were issued before the system was implemented.

Zuo et al. [9] mention a decentralized system for Renewable Energy Certificates (RECs) where the RECs are tokenized as non-fungible tokens (NFTs). Hence, to create atomicity, security, and reliability in certificates, if any certificate is invalid, it will not get passed into the network.

To overcome the above limitation, a blockchain-based certificate storage system called "Verificate"is proposed. In Verificate, the user can upload certificates and get them verified by the respective issuing authority. Only after successful verification, the uploaded certificates are stored in a decentralized manner. These certificates can then be accessed by the interviewer whenever needed.

Having delved into previous research work, let's now turn our attention to the technologies at play within our system. Below are the technologies that drive the methodology, offering a thorough explanation of their responsibilities in reaching our study objectives.

The proposed model uses Polygon, a side chain of the Ethereum blockchain, as our testnet because Polygon's Layer 2 scaling solutions closely mimic the real-world performance of a blockchain network. Furthermore, Web3.storage leverages the IPFS protocol to store files in a distributed way, aiming to conserve space while maintaining secure data storage. We adopted Hardhat as our development environment for the purposes of compiling and testing smart contracts.

Node.js, which is used to develop the backend of Verificate, is a server-side JavaScript runtime environment that is built on Chrome V8 and pairs with Express.js, a web framework of Node.js. It facilitates asynchronous programming and aids to handle connections efficiently. This helps when working on real-time applications and data heavy projects. Express.js enhances development by offering a structured framework for building web applications and APIs. There is a need for temporary database storage in the system; hence, MongoDB, an open-source, platform-independent database application, is used.

3 Proposed Model

Various systems claim to verify certificates and prove their authenticity. Most of these are centralized systems. Centralized systems are vulnerable to threats, such as hacking, which may compromise the authenticity of verified certificates. Unauthorized access to these systems poses a threat of tampering with the stored certificate files.

Our proposed system, Verificate, uses the concepts of decentralization and distributed storage to overcome the above-mentioned issues. The use of distributed storage ensures safer and tamper-proof storage of the certificate files. As a second layer to safeguard the certificate files and maintain trust in their authenticity, we use blockchain to save the storage details of the files and additional information about the certificates (Fig. 12.1).

Our system, Verificate, broadly has three stakeholders: the student/candidate, the verifier, and the recruiter.

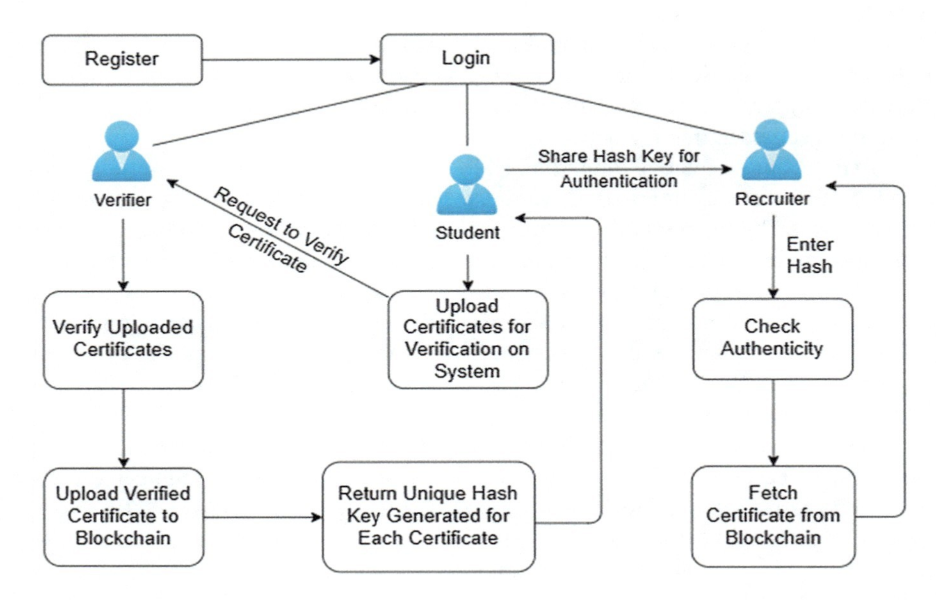

Fig. 12.1 Flow diagram of verificate

1. **Student/Candidate:** The student/candidate is the one who wants his/her certificate to be verified. A candidate is a certificate holder and will be uploading the certificate onto the system for verification and uploading it onto the blockchain.
2. **Verifier:** The Verifier is an external entity, an organization or university, which has issued the certificate. It is responsible for authenticating the certificate uploaded by the candidate. The certificate is uploaded onto the blockchain only if it is approved by the respective certificate-issuing university or organization.
3. **Recruiter:** A Recruiter (recruiting organization) can be an interviewer, a university or any other body that wants to check the authenticity of the certificate successfully uploaded onto our system by entering the unique hash given by the candidate into the Verificate system.

The overall process unfolds as follows: The student uploads their certificate onto the Verificate system. The system forwards the certificate to the verifier to check its legitimacy. Once the certificate receives approval from the verifier, Verificate uploads it onto the blockchain and returns a unique hash key to the student. The student can then share this hash key with the recruiter. To verify the certificate's authenticity, the recruiter enters the hash key into Verificate. If the hash is valid, Verificate returns the certificate to the recruiter. If not, then the recruiter is informed about the same.

4 Methodology

With Verificate, the system considers three stakeholders, the student/candidate, verifiers, and recruiters. Every stakeholder performs specific functions that contribute to the verification process. The process begins when a student uploads their certificates on the system. Initially, the certificate files get stored in the local file system in a dedicated location unique to each student. While the students upload their files, they are requested to mention the issuing authority for each certificate and additional information like description of certificate. This information helps the system to direct the verification request to the respective verifiers. On successful upload of the files, a data field gets associated with each file that is supposed to hold the hash value of the blockchain transaction if the certificate is declared authentic (Fig. 12.2).

When a student uploads the certificate, the respective verifiers can view the uploaded files to be verified. The authorities declare the authenticity of the certificate files by performing checks proposed by their organization. Verificate is not involved in this process, assuming every organization may use differing methods. Once the certificate is verified, the verifier marks it as genuine on the system.

The process of preserving the authenticity of the uploaded certificate file follows. Utilizing the distributed nature of Web3.storage, which uses IPFS protocol, the

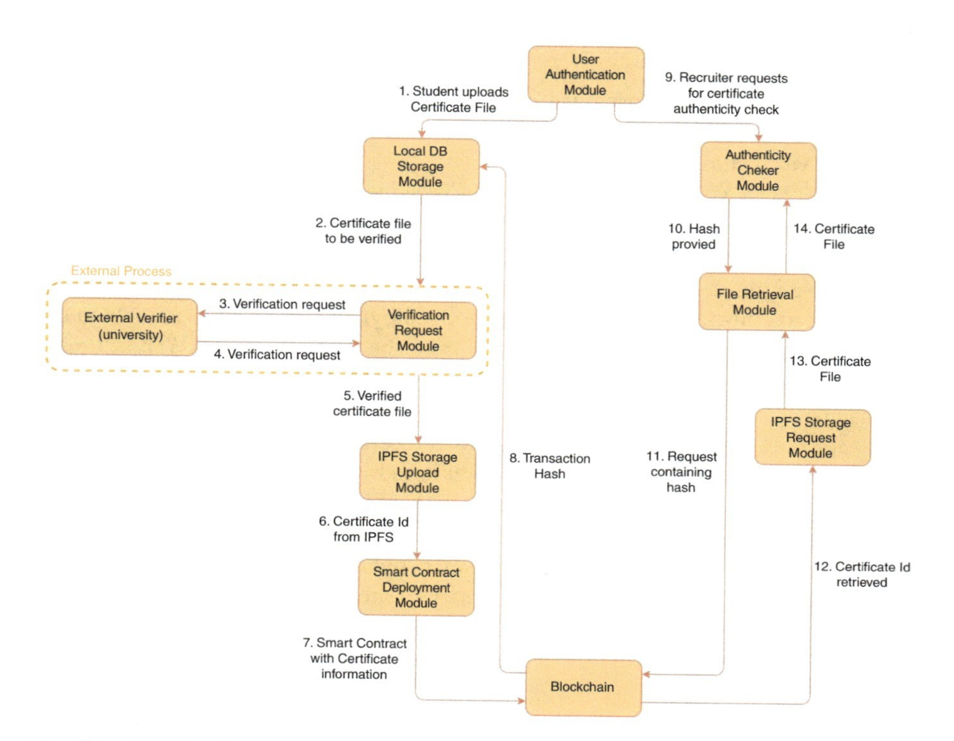

Fig. 12.2 Modular diagram of verificate

certificate files are uploaded to it. A unique value, called Content Identifier (CID), is returned for every uploaded file. The CID, student id, current timestamp, issuing institute name, and other necessary information for each file are stored in a Smart Contract and sent to the block creation process of the blockchain. The transaction hash returned is stored in the local database, created using MongoDB, and associated with the respective certificate. This hash value is provided to the student that they are supposed to share with their recruiter when necessary.

When the recruiter wishes to check the authenticity of the certificates provided by the candidate, they query the system with the certificate's hash. Our system fetches the Smart Contract associated with the hash value and retrieves the CID. If the system fails to find the Smart Contract using the provided hash, it alerts the recruiter that the certificate is unauthenticated. The CID is used to query Web3.storage API and the certificate file is displayed. The certificate is genuine if the system can fetch the certificate file from Web3.storage.

Using blockchain technology with the IPFS storage, the system establishes a two-layer security framework to safeguard the authenticity of the verified certificates. The stakeholders can benefit from our system with minimal effort and minor changes to their existing processes.

5 Implementation

To create Verificate, we used various industry-standard technologies. Here we describe how we use those technologies to make Verificate a robust and reliable certificate verification platform.

We used HTML, CSS, JavaScript, and Embedded JavaScript Templating (EJS) to create responsive and dynamic web pages. The server-side processing by sophisticated programs produced using Node.js handles user requests such as user authentication, file upload, verification, and certificate retrieval. It also performs process invocation like file storage to Web3.storage, deployment of Smart Contracts to the blockchain network, information retrieval from the blockchain and Web3.storage, and database queries. We store the information about the users and the certificates in a database created using MongoDB, a NoSQL database. The stored data are necessary for functions like user authentication and certificate verification.

We use Web3.storage, a storage solution based on IPFS protocols, to store the verified certificate files. Being distributed in nature, it ensures the safe storage of the files and acts as the first layer of defence in Verificate. We achieve the second layer of security using the Ethereum blockchain network to deploy Smart Contracts. Initially, we tested the deployment of Smart Contracts on the Goerli Testnet but soon discovered the gas cost to be very high. Therefore, we switched the implementation to Polygon's Mumbai Testnet as it offers faster deployment performance and lower gas cost.

We collect the test currency from faucets offered by providers such as Alchemy and store it in a MetaMask account that connects with Verificate to compensate for the funds required to perform the transactions on the blockchain. The Smart Contracts are built using Solidity and initially tested using the Remix IDE. Later, we used Hardhat for final testing and automated deployment of smart contracts to the blockchain. With the amalgamation of all the mentioned technologies, we establish a seamless experience of certificate verification for the users.

5.1 System Snapshots

Figure 12.3 shows the landing screen of the system. Users can view the information about the system and choose to either login or register from this page.

Figure 12.4 shows the dashboard wherein the students can see a graph that displays the statistics of the number of certificates uploaded and verified daily by system. They can also see all the certificates that they have uploaded for verification and their verification status.

Figure 12.5 shows the screen from which the students can upload their certificates for verification.

Figure 12.6 shows the page from which verifiers can view the certificate verification requests and update their verification status.

Figure 12.7 shows the screen where recruiters can enter the submitted hash key (by candidate) and check their authenticity.

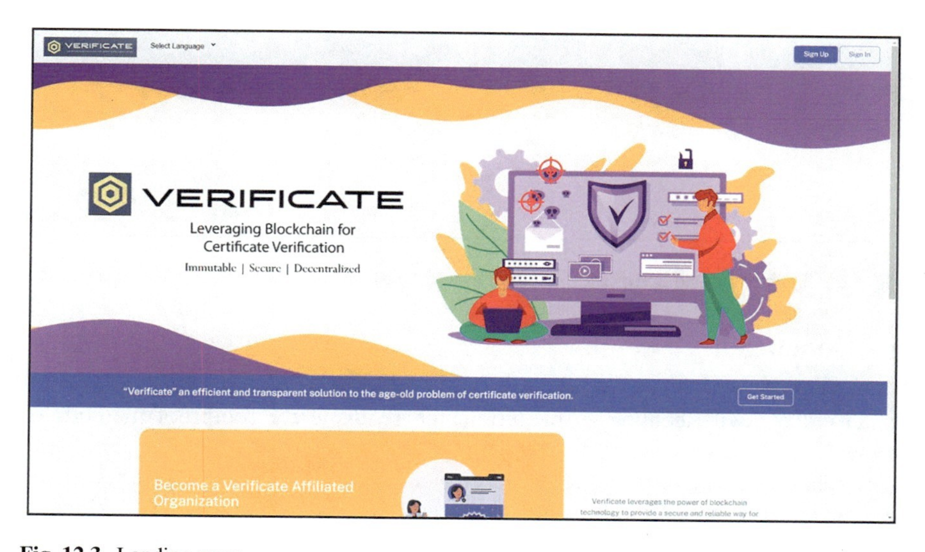

Fig. 12.3 Landing page

Fig. 12.4 Dashboard

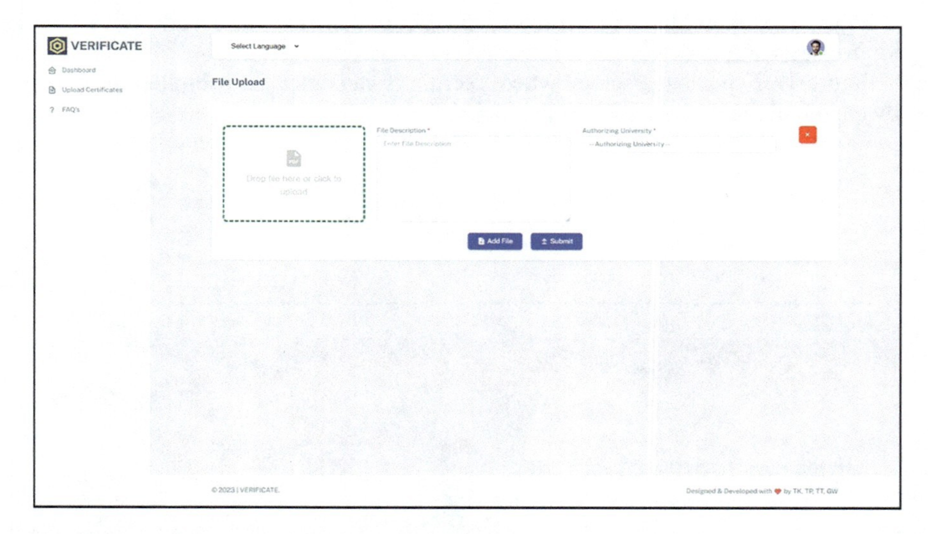

Fig. 12.5 Page to upload certificate file (student interface)

6 Conclusion

In an era of swift technology evolution, the requisite for foolproof methods of authenticating and safeguarding certificates has become vital. The escalating instances of deceptive alterations in certificate documents have exacerbated the urgency of robust solutions. With Verificate, we try to address this exigency,

Fig. 12.6 Certificate verification page (verifier interface)

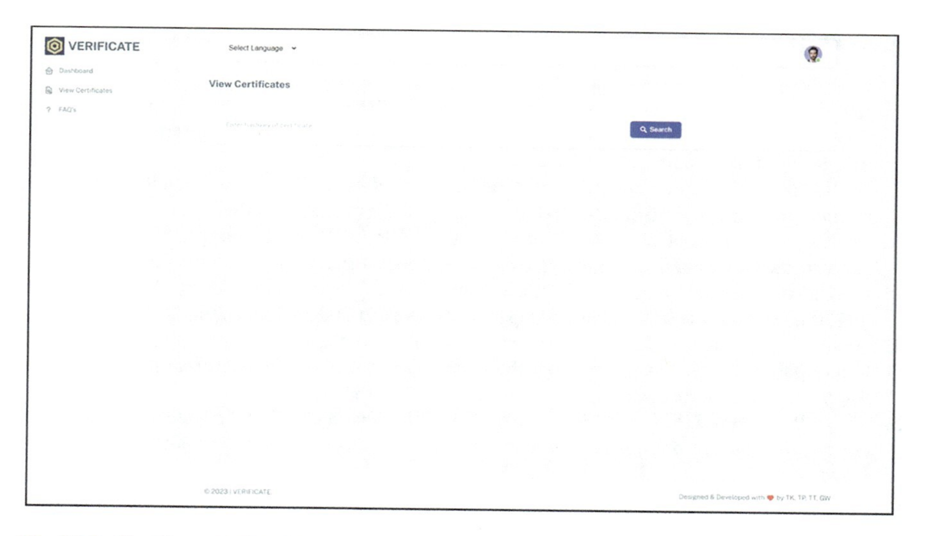

Fig. 12.7 Certificate retrieval page (recruiter interface)

propounding a method that can curtail the impact of counterfeit certificates on the community.

Our approach emerges as a comprehensive remedy that prevails over the limitations of existing methods. Using Blockchain technology, we harness the potency of decentralization to fortress against the rising tide of falsified credentials. With a unified point of access, Verificate brings cohesion to all stakeholders involved, alleviating the complexity and streamlining the verification process.

To demonstrate this, we performed a case study that displays the efficiency gain with Verificate over the traditional method. The values taken into consideration are rough estimates of the real world. The traditional method involves sending hard copies of certificates to the issuing authority. The one-way transit of these documents requires an average of 15 days. Assuming a timeframe of 15 days for the verification process by the verifiers, it adds up to approximately 45 days to complete the transaction. On the other hand, Verificate eliminates the transit time. We still assume a timeframe of 15 days for the verification of certificates by the verifiers. Therefore, the total time required for the verification process using Verificate effectively shortens to one-third of what it takes in the traditional approach.

In conclusion, Verificate represents a technical innovation with our commitment to shaping a future where integrity, authenticity, and fairness take centre stage. As a tangible step in this direction, one enhancement we contemplate involves the integration of an Optical Character Recognition (OCR) module. This module could automatically identify the issuing authority, facilitating a harmonious and error-free transmission of documents to relevant verifiers.

References

1. All India Survey on Higher Education (AISHE) 2020–2021. https://pib.gov.in/PressReleasePage.aspx?PRID=1894517#:~:text=The%20total%20number%20of%20pass,94%20Lakh%20in%202019%2D20
2. INDIA 2030: Vision of assessments in creating a skilled nation. https://community.nasscom.in/communities/talent-skills/india-2030-vision-assessments-creating-skilled-nation
3. Trends in Education Verification, AUTHBRIDGE-2020. https://authbridge.com/resources/nearly-28-of-job-applicants-submit-fake-degrees-says-annual-trend-report-2020/#:~:text=Trends%20in%20Education%20Verification&text=According%20to%20the%20recent%20Annual,discrepancies%20in%202020%20in%20India
4. Gupta, D., Chaubey, S., Ram, A., Raheman, A., & Ratnamala, A. (2022). Mapping & visualization of education systems using blockchain technology. *International Journal for Research in Applied Science & Engineering Technology, 6*, 1128–1131.
5. Shuyi, P., & Lam, J. S. L. (2022). *The benefits of blockchain for digital certificates: A multiple case study analysis*. School of Civil and Environmental Engineering, Nanyang Technological University, 50 Nanyang Avenue, 639798.
6. Lamkoti, R. S., Shetty, H., & Prof. Bharati Gondhalekar. (2021). Certificate verification using Blockchain and generation of transcripts. *International Journal of Engineering Research & Technology (IJERT), 10*.
7. Gayathri, A., Jayachitra, J., & Dr. S. Matilda. (2020). *Certificate validation using blockchain*. IEEE 7th International Conference on Smart Structures and Systems ICSSS.
8. Nyaletey, E., Parizi, R. M., Zhang, Q., & Choo, K.-K. R. (2019). *Block IPFS – Blockchain-enabled interplanetary file system for forensic and trusted data traceability*. IEEE International Conference on Blockchain (Blockchain).
9. Zuo, Y. (2022). *Tokenizing renewable energy certificates (RECs)—A blockchain approach for REC issuance and trading* (Vol. 10, p. 134477). Department of Accountancy and Information Systems, University of North Dakota.

Chapter 13
Transforming Waste Management Practices Through Blockchain Innovations

Ritu Vats and Reeta

1 Introduction

In the modern world, waste management is one of the important challenges that has to be addressed. Waste management is connected to other global issues including sustainable production and consumption, food and resource security, poverty alleviation, and climate change. Governmental organisations, non-governmental organisations, judicial institutions, and technological firms all have a role to play in addressing the issues facing the waste management sector. With the correct combination of all stakeholders cooperating on the same platform, waste management might be managed, according to a number of study reports and our own real-world experiences. Before anybody attempts to develop technology-backed solutions, it is crucial to understand the waste management life cycle and the multiple obstacles faced at each stage. The United Nations (UN) defines waste management as the activities of (a) collecting, moving, treating, and disposing of waste; (b) controlling, monitoring, and regulating the production, moving, treating, and disposing of waste; and (c) preventing waste production through in-process modifications, reuse, and recycling.

Figure 13.1 illustrates the numerous processes garbage travels through before it is disposed of at a landfill.

Waste management is everyone's duty who creates garbage, not just one person or business. We list some of the most significant issues that need to be resolved below.

R. Vats (✉) · Reeta
Department of Management, Maharishi Markandeshwar (Deemed to be University), Mullana, Ambala, Haryana, India

S. M. Idrees, M. Nowostawski (eds.), *Blockchain Transformations*, Signals and Communication Technology, https://doi.org/10.1007/978-3-031-49593-9_13

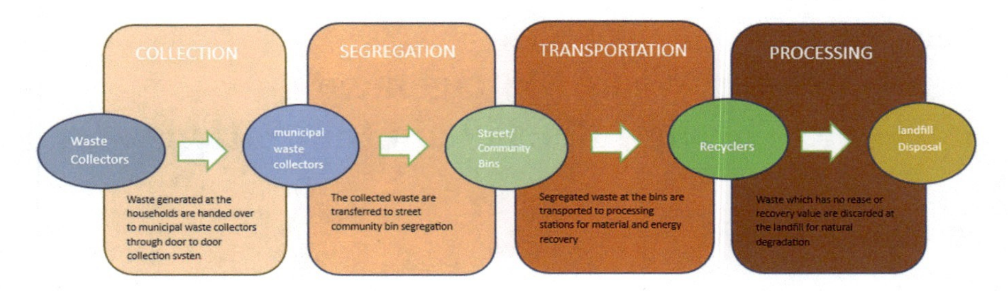

Fig. 13.1 Waste management cycle

1.1 Failure to Enforce Laws and Raise Awareness

Inadequate waste management has a number of causes, but one of the biggest ones is ignorance. People who produce garbage are not aware of the need for and methods of waste segregation, and waste handlers and collectors are not aware of the proper methods for processing and disposing of both usable and non-useful trash. The effects of poor waste management on health and the environment are often not well known. According to India's 2016 waste management regulations, source segregation of waste is required in order to channel waste to wealth through recovery, reuse, and recycling.

1.2 An Absence of Source-Level Waste Segregation

The primary producers of trash are families and businesses, where it is necessary to separate garbage at the source. The majority of garbage currently being collected by communities or municipalities is not effectively separated into recycleable and non-recycleable waste. In a few instances, the government fails to provide the general communal with the necessary infrastructure for the disposal of separated garbage. Government regulations state that trash generators must pay a "User Fee" to the garbage collector and a "Spot Fine" for littering and non-segregation; however, compliance with the regulations is extremely minimal.

1.3 Abundant Generation

Innumerable methods contribute to the daily growth of garbage being produced.

The amount of waste produced in a country varies with its per-capita wealth. Population growth and garbage production are both rising in low- and middle-income nations. In 2016, 62 MT of waste were produced annually in India, of which 5.6 MT were plastics wastes, 0.17 MT were biomedical wastes, 7.90 MT were hazardous wastes, and 1.5 MT were electronic wastes, according to statistics released by the Ministry of Human Resource Development, Government of India.

Additionally, Indian cities produce 200–600 g of garbage per person each day. Actually, only around 75–80% of municipal garbage is collected, and only 22–28% of this waste is processed and treated. Of the 43 MT per year that are collected, only 11.9 MT are treated and the other 31 MT are disposed of in landfills. In 2030, trash output is anticipated to rise from 62 MT to over 165 MT.

1.4 Shortage of Scientific Landfills

Due to the contamination of groundwater and the release of methane gas into the atmosphere, landfills built nowadays are not created scientifically. Other problems with waste sites include offensive odours, fire explosions, and animal scavenging. The quantity of dump sites that are readily accessible does not match the demand in the nation. Over 70% of the collected municipal garbage is just deposited into landfills.

1.5 Inadequate Use of Technology to Monitor Waste Flow

Waste producers and handlers are unaware of how and where their garbage is treated once it has left their bin. There is no reliable waste management tracking system. Although RFID technology is utilised for trash management, its application is not widespread. Data input errors, inconsistent data entry, and data manipulation for financial advantage might result from the present waste management system's human data entering. Additionally, official correspondence is transmitted on paper, which increases the risk of loss in transit.

1.6 Lack of Responsibility

Once the garbage has left the creators' control, they never again accept responsibility for it. Because of this, there is an excessive amount of rubbish left behind in the streets and communal areas.

Cryptographic ledger is a decentralised, distributed, immutable, computing, and information-sharing platform that enables several authority domains—which do not trust one another—to interact, coordinate, and collaborate in a logical decision-making process. Cryptographic ledger is the enabling technology that made the "Satoshi Nakamoto"-created Bitcoin cryptocurrency possible. Cryptographic ledger is a continuously expanding collection of blocks with timestamps linked together by cryptographic hashes. Because blocks may only be added to this chain, the

Cryptographic ledger is unchangeable. High computational work and approval from other network nodes are needed to edit a block, which is essentially impossible. Computational nodes, not by a single person, but by a network. The transparency of the Cryptographic ledger is a result of the information being shared and accessible to anyone. With a wide range of applications, including but not limited to banking, healthcare, government, manufacturing, and distribution, Cryptographic ledger has been recognised to have greater promise than merely as a platform for cryptocurrencies. Supply chain management, startup funding, electronic voting, educational credentialing, electricity generation, and distribution are a few examples of the many fields where Cryptographic ledger is being used.

The chapter is set up as follows:

The Cryptographic ledger-based waste management solutions are covered in Sect. 2 of this chapter. We have examined the current solutions in part 3 before analysing the future work in Sect. 4.

2 Waste Management Using Cryptographic Ledger

This section will cover the current Cryptographic ledger-based waste management systems.

2.1 Swachh Coin

Swachh coin is a Cryptographic ledger-based method for micromanaging garbage, especially from homes and businesses, and turning them into valuable goods in an effective and environmentally responsible way. Among the high-value products produced from the processed wastes are electricity, paper, steel, wood, precious metals, glass, and polymers. The Swachh Ecosystem is a Decentralised Autonomous Organisation (DAO), which is managed independently based on set rules in the form of smart contracts.

Swachh coin implements an iterative process cycle using a number of cutting-edge technologies, and over time, this will entirely autonomise, optimise, and produce the system. This cycle of repetitive processes concentrates on the data shared among many ecosystem participants, analyses those data, and offers ideas in real time based on predictive techniques.

The technologies and tools that make up the Swachh coin ecosystem are listed below.

1. SwATA (Swachh Big Data): One of the issues facing the waste management sector is data management and openness with tonnes of garbage being produced, gathered, moved, processed, and then disposed of. A sufficient amount of data is evenly generated. SwATA is a tailored application that gathers, saves, and evalu-

ates this data in order to offer recommendations for different improvement tasks including route optimisation, maintenance cycles, and report production. In order to create highly organised data that the SWATA application can analyse, SWATA employs a NoSQL methodology and virtual data filters at the collection points. The data created in an area like waste management is unstructured. The most cutting-edge and trustworthy technique used by SwATA is prescriptive analysis. The immutability of the data is provided by Cryptographic ledger technology.

2. SwATEL (Swachh Adaptive Intelligence): SwATEL is referred to as the brain of the whole ecosystem since it allows different types of machinery and equipment in the ecosystem to communicate and coordinate with one another, giving them intelligence. In order to make judgements based on prior learning in real time, SwATEL utilises a customised application of adaptive intelligence (AI). These choices may result in physical or digital acts that are then recorded on the Cryptographic ledger. Deep learning and neural networks are the two main components of AI.
3. SwIOT (Swachh Internet of Things): The Internet of Things (IoT) enables us to manage everything that is online. The collection and transportation trucks, collection bins, treatment facilities, and disposal sites might all be connected to and managed by an IoT-based network in the waste management industry.
4. SwBIN (Swachh Bins): Correct and thorough garbage collection will improve the waste management procedure.

   SwBIN is similar to our standard trash cans but has more modern, appealing features like decentralised advertising, automated lid shutting and opening, and free Wi-Fi. A Unique Identifier (UID) is given to each waste producer. When waste is dumped, SwBIN will recognise the user using the QR code that corresponds to the UID and measure factors such as the quantity and quality of the garbage to determine the rewards points. The Cryptographic ledger contains a record of these points. The users will receive this prize in the form of Swachh Tokens. The status of the trash in the bin will be sent to the waste service providers via SwBIN via the SwIOT. The SwBIN installation costs throughout all regions are to be covered by the advertisement, according to the Swachh Foundation's proposal. The garbage collecting and processing facilities receive the collected wastes.
5. Cryptographic ledger and Smart Contracts: Swachh coin chose Ethereum as their Cryptographic ledger solution with smart contracts'support, creating a Decentralised Autonomous Organisation that fits their present process and requirements. Autonomous Philanthropy is made possible by Swachh coin's DAO Smart contract.
6. Swachh Tokens: Swachh Tokens (SCX) are utility tokens that a user may earn as compensation by using correct waste disposal techniques. Platform-specific settlements might be the main application for these tokens. In addition to money, these tokens give their owners the ability to nominate NGOs for funding or cast a vote on important platform choices. The 400 million Swachh tokens that the Swachh coin network intends to develop were distributed in 69% crowd sales.

2.2 Recereum

A Cryptographic ledger network called Recereum allows garbage and recyclables to be converted into actual currency. The garbage collection company and individual consumers may now communicate directly thanks to the Cryptographic ledger. The Recereum Cryptographic ledger compensates individual home users with Recereum coins, which are generated from the money that is saved by properly sorting garbage. The Recereum ecosystem runs on Ethereum, the biggest communal Cryptographic ledger. The transfer of rewards (tokens) from one account to another is recorded on the Cryptographic ledger. According to the Recereum whitepaper, supply chain management, payments, and smart contracts are the major uses. Recereum tokens (RCR) [1] are Ethereum platform-issued ERC20-based coins. Recereum will issue 7,999,000 RCR, of which 65% will be sold to the general communal, at a pricing of 1 ETH for 300 RCR.A vending machine or battery collecting device might be used in conjunction with the Recereum Cryptographic ledger technology for garbage collection. Recereum focuses primarily on the cycle of waste management‘s trash sorting and associated sectors.

2.3 The Plastics Bank

With the goal of stopping the flow of plastics into the ocean by making people money, Plastics Bank is a Cryptographic ledger-based application. This initiative's plastics collection is recycled, and the resulting Social Plastics is then sold. These plastics were approved by Plastics Bank and offered the collector a premium as a reward. The safest and most reliable way to create a social effect that can scale throughout the globe is by leveraging Cryptographic ledger technology to distribute, authenticate, and store these awards (Fig. 13.2).

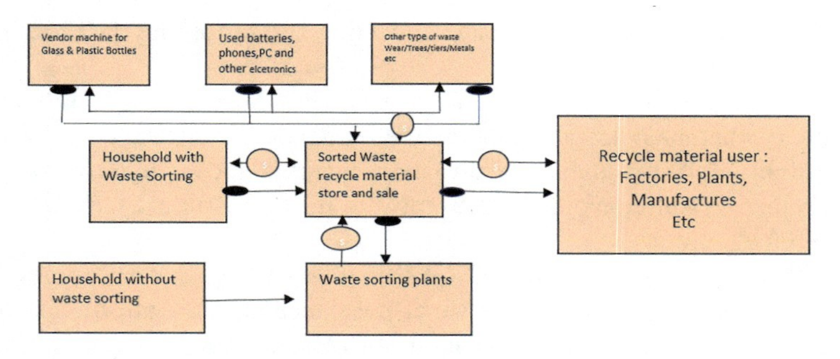

Fig. 13.2 Recereum Cryptographic ledger workflow

3 Analysis and Observations

The communal Ethereum Cryptographic ledger provides the foundation for the majority of the existing Cryptographic ledger solutions outlined in Sect. 2. Some of the observations are listed below:

1. Swachh coin is built on a Decentralised Autonomous Organisation (DAO) that uses smart contracts to function. However, the DAO itself allows attackers to steal money from the DAO due to flaws in the smart contracts. Ethereum forked as a result of this. The Cryptographic ledger technology and the smart contracts that operate on it are still in the development stage and cannot yet address practical problems like waste management. It is crucial at this point to refrain from having too high of an idealistic expectation that Cryptographic ledger technology would be able to solve every problem.
2. The previously proposed Cryptographic ledger solution: We recognise that the IoT-based system creates a large amount of data, which should also be kept on the Cryptographic ledger. At the time of writing, there is no viable consensus method for IoT-based networks operating in real time, despite the fact that some use cases in the waste management area are not time essential.
3. Furthermore, present solutions do not address all of the issues associated with waste management. The Cryptographic ledger-based system delivers accountability, openness, and transparency in the process and data management, data security through the immutable property of Cryptographic ledger, incentivisation through cryptocurrency for waste segregation, and awareness. However, it does not solve the law enforcement and non-compliance issues that are prevalent in waste management.

4 Conclusion and Future Work

However, the scope of the current research is rather broad since it not only discusses the triple bottom-line components of sustainability but also compiles a list of publications pertinent to the supply chain for humanitarian aid. However, the study has certain drawbacks, such as the fact that it only looked at sustainability and did not include other supply chain characteristics like resilience, agility, or robustness. In this study, we explored the waste management process and the issues encountered along the waste management life cycle. We have described Cryptographic ledger technology and emphasised the benefits of adopting Cryptographic ledger to address these difficulties. We also analysed significant current Cryptographic ledger-based waste management systems such as Swachh coin, Recereum, and Plastics Bank and stated our findings based on our investigation. As a future project, we propose developing Thui-mychain, a Cryptographic ledger solution for waste management based on IoT, AI, and our native cryptocurrency Naanayam, which would incorporate the United Nations goals for waste management and address the challenges identified above.

Reference

1. Recereum Coins [Online]. Available: https://icobench.com/ico/recereum

Chapter 14
Decentralized Technology and Blockchain in Healthcare Administration

Anamika Tiwari, Alisha Sikri, Vikas Sagar, and Roshan Jameel

1 Introduction

In recent years, decentralized technology and blockchain have emerged as transformative tools with the potential to revolutionize various industries, including healthcare administration. This chapter explores the applications, benefits, and challenges of integrating decentralized technology and blockchain into healthcare administration processes.

1.1 Blockchain Technology

Blockchain technology is a decentralized and distributed digital ledger system that allows multiple parties to record and maintain a secure and transparent record of transactions. It was originally created to support the digital crypto currency Bit coin, but its potential applications extend far beyond just financial transactions [1]. At its core, a blockchain is a chain of blocks, where each block contains a list of transactions. These blocks are linked together in a chronological order, forming a continuous and unchangeable chain [5]. The key features of blockchain technology include:

A. **Decentralization**: Unlike traditional centralized systems, where a single entity has control over data and transactions, a blockchain operates in a decentralized manner. It is maintained by a network of participants (nodes) rather than a single central authority.

A. Tiwari (✉) · A. Sikri · V. Sagar · R. Jameel
Noida Institute of Engineering and Technology, Noida, India

S. M. Idrees, M. Nowostawski (eds.), *Blockchain Transformations*, Signals and Communication Technology, https://doi.org/10.1007/978-3-031-49593-9_14

B. **Transparency**: Transactions recorded on a blockchain are visible to all participants within the network. This transparency helps prevent fraud and ensures that everyone has access to the same information [2].
C. **Immutability**: Once a transaction is recorded in a block and added to the blockchain, it becomes extremely difficult to alter or delete. This is achieved through cryptographic hashing and consensus mechanisms.
D. **Security:** Blockchain uses advanced cryptographic techniques to ensure the security of transactions and data. Transactions are verified and added to the blockchain through consensus mechanisms, which prevent unauthorized changes [6].
E. **Consensus Mechanisms**: These are protocols that ensure that all participants in the network agree on the state of the blockchain. Different consensus mechanisms (e.g., Proof of Work, Proof of Stake) have been developed to achieve agreement among participants in various ways [3, 4].
F. **Smart Contracts**: These are self-executing contracts with the terms of the agreement directly written into code. They automatically execute and enforce the terms when predefined conditions are met, eliminating the need for intermediaries [7].

Blockchain technology has numerous potential applications beyond crypto currencies, including supply chain management, identity verification, voting systems, intellectual property protection, healthcare data management, and more. Its ability to provide trust, security, and transparency in various industries has led to significant interest and ongoing research in its potential applications [2]. It is important to note that while blockchain technology offers many advantages, it is not without limitations, including scalability issues, energy consumption concerns (especially in Proof of Work systems), and the complexity of implementing certain applications. Different blockchain platforms and variations continue to be developed to address these challenges and tailor the technology to specific use cases (Fig. 14.1).

(a) **Decentralized Technology in Healthcare:** Decentralized technology characterized by its distributed nature and lack of central authority holds promise in healthcare administration. Through the use of decentralized networks, data can be securely stored, accessed, and shared across multiple parties, enhancing data integrity and reducing the risk of single points of failure.
(b) **Blockchain and Healthcare Data Management:** Blockchain technology, a type of decentralized ledger, offers a transparent and tamper-resistant way to manage healthcare data. Patient records, medical histories, and other sensitive information can be securely stored on the blockchain, granting patients control over their data and allowing healthcare providers seamless access to accurate information [10].
(c) **Improving Interoperability and Data Exchange:** One of the key challenges in healthcare administration is the lack of interoperability among different systems and providers. Blockchain's standardized protocols can facilitate seamless data exchange between various stakeholders, leading to more efficient care coordination and reduced administrative burden.

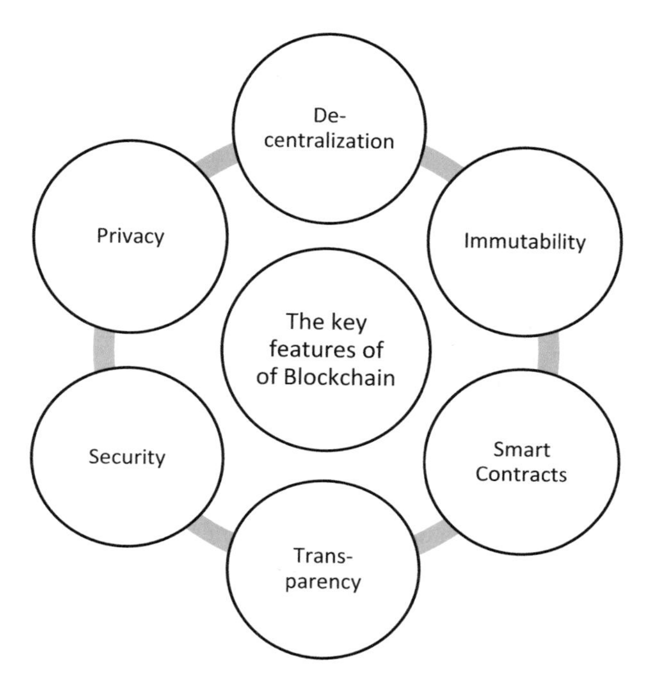

Fig. 14.1 The key features of blockchain technology

(d) **Enhancing Patient Privacy and Security:** Blockchain's cryptographic principles enhance patient privacy by allowing them to share only necessary parts of their medical history while keeping the rest of their data confidential. Additionally, blockchain's immutable nature helps prevent unauthorized access and tampering of medical records, contributing to higher data security [2, 3].

(e) **Smart Contracts for Administrative Efficiency:** Smart contracts, self-executing agreements triggered by predefined conditions, have the potential to streamline administrative processes in healthcare. These contracts can automate tasks such as insurance claims processing, billing, and supply chain management, reducing errors and saving time.

2 Decentralized Blockchain Features and Implementation in Healthcare

Decentralized blockchain technology has the potential to address various challenges and improve different aspects of the healthcare industry. Here are some features and potential implementations of decentralized blockchain in healthcare:

(a) **Data Interoperability and Sharing**:
Blockchain can enable secure and interoperable sharing of medical data across different healthcare providers, systems, and institutions. Patients could have control over their health records and grant permission for specific entities to access their data.
(b) **Data Security and Privacy:**
Blockchain's cryptographic mechanisms can enhance the security and privacy of patient data. Personal health information could be stored in a distributed manner, reducing the risk of single points of failure and unauthorized access.
(c) **Identity Management**:
Blockchain can be used to create tamper-proof digital identities for patients, healthcare professionals, and organizations. This could streamline identity verification processes and reduce the risk of fraud.
(d) **Clinical Trials and Research:**
Blockchain can facilitate the tracking and auditing of clinical trial data, ensuring transparency and reducing the chances of data manipulation. Researchers can have access to trustworthy data while maintaining patient privacy.
(e) **Drug Traceability and Supply Chain Management**:
Blockchain can improve the transparency and traceability of pharmaceutical supply chains, reducing the prevalence of counterfeit drugs and ensuring the authenticity of medications from manufacturer to patient.
(f) **Telemedicine and Remote Patient Monitoring**:
Blockchain could enable secure and verifiable storage of telemedicine interactions and remote patient monitoring data. This can help ensure the integrity of the data and maintain patient privacy [8].
(g) **Billing and Claims Management**:
Blockchain can streamline billing and claims processes by creating transparent and auditable records of medical services provided and billed. This can reduce errors and fraud in billing.
(h) **Research Data Sharing:**
Researchers and institutions can securely share research data with each other using blockchain. This would encourage collaboration, data sharing, and replication of research results.
(i) **Health Insurance and Verification:**
Blockchain can facilitate verification of insurance coverage and streamline the claims process. This could lead to faster and more accurate claims settlement.

3 Implementing Blockchain in Healthcare

Regulations and Compliance: Healthcare is heavily regulated, and any blockchain implementation must adhere to relevant laws, such as data privacy regulations (e.g., HIPAA).

- Interoperability: To achieve data sharing and integration, blockchain solutions should be designed to work with existing healthcare systems and standards.
- Scalability: Blockchain networks need to handle a significant amount of data, especially in healthcare. Scalability solutions need to be considered.
- Adoption: Adoption by healthcare providers, organizations, and patients is crucial for the success of any blockchain solution. User-friendly interfaces and education are essential.
- Security and Privacy: While blockchain enhances security, it is not immune to all threats. Robust security measures must be in place to protect sensitive healthcare data [11].
- Collaboration: The implementation of blockchain in healthcare may require collaboration among stakeholders, including healthcare providers, technology companies, regulators, and patients.

Overall, decentralized blockchain technology has the potential to revolutionize healthcare by addressing data security, interoperability, transparency, and patient-centricity. However, its successful implementation requires careful planning, collaboration, and consideration of the unique challenges within the healthcare industry [12] (Table 14.1).

- Challenges and Considerations: While the potential benefits are substantial, integrating decentralized technology and blockchain into healthcare administration presents challenges. These include regulatory compliance, scalability, energy consumption, and the need for industry-wide standards.
- Real-World Applications: Several real-world examples showcase the impact of decentralized technology and blockchain in healthcare administration. Projects like MedRec and Medical chain are leveraging these technologies to enable

Table 14.1 Decentralized blockchain features and implementation in healthcare

Application	Description
Decentralized Technology in Healthcare Decentralized technology	Accurate identification and localization of abnormalities, tumors, and diseases through image segmentation, classification, and feature extraction
Blockchain and Healthcare Data Management	Quantitative measurement of disease progression by analyzing changes in tissue density, blood flow, or metabolic activity over time
Improving Interoperability and Data Exchange	Precise treatment planning based on the analysis of medical images to determine optimal strategies and assess potential outcomes
Research and Development	Investigation of new diagnostic methods, evaluation of treatment efficacy, and exploration of disease mechanisms through the analysis of large-scale medical imaging datasets.
Surgical Navigation	Image-guided surgical procedures using real-time image processing and analysis techniques to aid surgeons in precise interventions
Telemedicine	Remote analysis of medical images, enabling access to expert opinions and diagnostics in underserved areas

secure and efficient health data management [15], while others explore drug supply chain verification and telemedicine solutions.

4 Challenges in Integrating Decentralized Blockchain in Healthcare Administration

Certainly, here are some of the key challenges in integrating decentralized blockchain technology into healthcare administration:

1. **Regulatory Compliance**: Healthcare is a highly regulated industry, and introducing blockchain technology requires adherence to complex regulatory frameworks. Ensuring compliance with data protection laws such as HIPAA (Health Insurance Portability and Accountability Act) in the United States and GDPR (General Data Protection Regulation) in Europe is crucial. Navigating these regulations while maintaining the decentralized and transparent nature of blockchain poses a significant challenge [13].
2. **Data Privacy and Security**: While blockchain is known for its security features, managing patient data on a blockchain network requires careful consideration [14]. Striking a balance between data accessibility for healthcare providers and patient privacy is challenging. Ensuring that sensitive patient information remains confidential while being accessible to authorized personnel is a critical concern.
3. **Scalability**: The scalability of blockchain networks is an ongoing challenge. As healthcare administration involves vast amounts of data and numerous transactions, scaling blockchain to handle the high volume of data and maintain performance efficiency is a considerable hurdle. Solutions like shading and layer 2 scaling are being explored to address this issue.
4. **Interoperability**: Healthcare involves various stakeholders, including hospitals, clinics, pharmacies, insurers, and patients, often using different systems. Ensuring interoperability between these disparate systems and standardizing data formats for seamless data exchange is complex. Blockchain's potential to improve interoperability needs careful planning and implementation.
5. **Energy Consumption**: Many blockchain networks, especially those using proof-of-work (PoW) consensus mechanisms like Bit coin and Ethereum, are criticized for their high energy consumption. Sustainable blockchain solutions are sought to reduce the environmental impact while maintaining the network's security.
6. **Adoption and Education**: The healthcare industry is traditionally risk-averse, and introducing new technologies like blockchain requires buy-in from stakeholders. There is a need for education and awareness about the benefits of blockchain to encourage adoption [16]. Healthcare professionals and administrators must understand how blockchain works and how it can benefit their operations.

7. **Integration with Legacy Systems: Healthcare organizations** often have legacy systems in place. Integrating blockchain with these existing systems can be challenging and costly. Ensuring a smooth transition and compatibility with legacy infrastructure is essential.
8. **Costs and Investme**nt: Implementing blockchain technology in healthcare administration requires a significant initial investment in infrastructure, training, and ongoing maintenance. Healthcare organizations need to assess the cost-benefit ratio and long-term ROI to justify these investments.
9. **Data Migration and Legacy Data**: Transferring historical patient data from legacy systems to a blockchain can be a complex and time-consuming process. Ensuring the accuracy and integrity of this data during migration is essential for patient care and compliance.
10. **Ethical and Legal Issues**: Blockchain raises ethical and legal questions, such as the right to be forgotten (data erasure) and the ownership of healthcare records. Resolving these issues while maintaining transparency and trust in the blockchain network poses ethical and legal challenges. While blockchain technology has the potential to address many of the issues in healthcare administration, its integration is not without challenges. Overcoming these hurdles requires careful planning, collaboration between stakeholders, and a deep understanding of both the technology and the healthcare industry's unique requirements (Fig. 14.2).

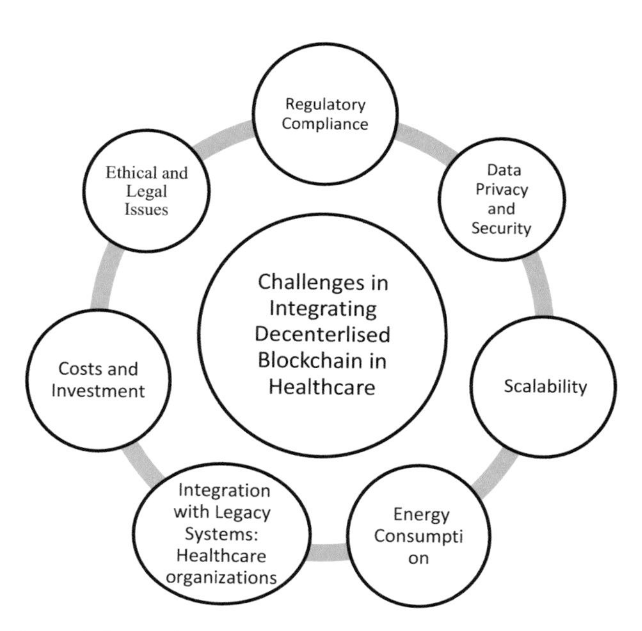

Fig. 14.2 Challenges in integrating decentralized blockchain in healthcare administration

5 Conclusion and Discussion

In conclusion, the integration of decentralized technology and blockchain in healthcare administration represents a promising step towards transforming the landscape of healthcare management. The potential benefits of these technologies, including enhanced data security, improved interoperability, streamlined administrative processes, and patient-centric care, hold the promise of revolutionizing how healthcare systems operate. Decentralized technology's distributed nature offers a solution to challenges like data silos and interoperability issues that have plagued the healthcare industry for years. By providing a secure and transparent platform for data exchange, blockchain can enable seamless sharing of patient information while maintaining patient privacy and consent.

The application of blockchain in healthcare administration extends beyond data management. Smart contracts can automate and streamline administrative tasks, reducing errors and accelerating processes such as billing, claims processing, and supply chain management. This automation not only saves time and resources but also reduces the potential for human error. However, it is important to acknowledge that the integration of these technologies is not without its challenges. Regulatory compliance, data privacy, scalability, energy efficiency, and the cost of implementation all pose significant hurdles that must be carefully navigated. The healthcare industry's cautious nature and reliance on legacy systems add another layer of complexity to the adoption process.

Successful implementation requires collaboration between healthcare professionals, technology experts, regulators, and patients. Stakeholders must work together to design solutions that prioritize patient well-being, security, and compliance with legal and ethical standards.

In the coming years, as decentralized technology and blockchain mature, we anticipate witnessing more innovative use cases and solutions in healthcare administration [9]. While the road ahead may be challenging, the potential to create a more efficient, transparent, and patient-centric healthcare ecosystem makes the journey worthwhile. As we navigate these challenges and capitalize on the opportunities, the integration of decentralized technology and blockchain could play a pivotal role in shaping the future of healthcare administration. Decentralized technology and blockchain hold immense promise for transforming healthcare administration. By enhancing data security, interoperability, and administrative efficiency, these technologies pave the way for a more patient-centered, transparent, and effective healthcare ecosystem.

References

1. Zheng, Z., Xie, S., Dai, H., Chen, X., & Wang, H. (2018). Blockchain challenges and opportunities: A survey. *International Journal of Web and Grid Services, 14*, 352.
2. Harleen Kaur, M., Alam, A., & Jameel, R. (2018). A proposed solution and future direction for Blockchain-based heterogeneous medicare data in cloud environment. *Journal of Medical Systems, 42*, 156.

3. Suhasini, M., & Singh, D. (2021). Blockchain based framework for secure data management in healthcare information systems. *Annals of The Romanian Society for Cell Biology, 25*, 16933–16946.
4. Idrees, S. M., Nowostawski, M., & Jameel, R. (2021). Blockchain-based digital contact tracing apps for COVID-19 pandemic management: Issues, challenges, solutions, and future directions. *JMIR Medical Informatics, 9*(2), e25245.
5. Idrees, S., & Nowostawski, M. (2022). *Transformations through Blockchain technology*. Springer.
6. Idrees, S. M., Aijaz, I., Agarwa, P., & Jameel, R. (2021). Blockchain-based smart and secure healthcare system. In *Transforming cybersecurity solutions using Blockchain*. https://doi.org/10.1007/978-981-33-6858-3_9
7. Hathaliya, J. J., Tanwar, S., Tyagi, S., & Kumar, N. (2019). Securing electronics healthcare records in healthcare 4.0: A biometric-based approach. *Computers and Electrical Engineering, 76*, 398–410.
8. Faujdar, D. S., Sahay, S., Singh, T., Jindal, H., & Kumar, R. (2019). Public health information systems for primary health care in India: A situational analysis study. *Journal of Family Medicine and Primary Care, 8*, 3640–3646.
9. Liang, X., Zhao, J., Shetty, S., Liu, J., & Li, D. (2017). Integrating blockchain for data sharing and collaboration in mobile healthcare applications. In *Proceedings of the IEEE 28th annual international symposium on personal, indoor, and Mobile radio communications (PIMRC), Montreal, QC, Canada, 8–13 October 2017* (pp. 1–5).
10. Gupta, R., Thakker, U., Tanwar, S., Obaidat, M., & Hsiao, K. F. (2020). BITS: A Blockchain-driven intelligent scheme for Telesurgery system. In *Proceedings of the 2020 international conference on computer, information and telecommunication systems (CITS), Hangzhou, China, 5–7 October 2020* (pp. 1–5).
11. Hathaliya, J. J., Sharma, P., Tanwar, S., & Gupta, R. (2019). Blockchain-based remote patient monitoring in healthcare 4.0. In *Proceedings of the IEEE 9th international conference on advanced computing (IACC), Tiruchirappalli, India, 13–14 December 2019* (pp. 87–91).
12. Dinh, T. T. A., Liu, R., Zhang, M., Chen, G., Ooi, B. C., & Wang, J. (2018). Untangling Blockchain: A data processing view of Blockchain systems. *IEEE Transactions on Knowledge and Data Engineering, 30*, 1366–1385.
13. Reyna, A., Martín, C., Chen, J., Soler, E., & Díaz, M. (2018). On blockchain and its integration with IoT. Challenges and opportunities. *Future Generation Computer Systems, 88*, 173–190.
14. Alladi, T., Chamola, V., Parizi, R. M., & Choo, K. (2019). Blockchain applications for industry 4.0 and industrial IoT: A review. *IEEE Access, 7*, 176935–176951.
15. Rehman, E., Khan, M. A., Soomro, T. R., Taleb, N., Afifi, M. A., & Ghazal, T. M. (2021). Using blockchain to ensure trust between donor agencies and ngos in under-developed countries. *Computers, 10*, 98. 8. Ghayvat, H.; Pandya, S.; Bhattacharya, P.; Zuhair, M.; Rashid, M.; Hakak, S.; Dev, K. CP-BDHCA: Blockchain-b.
16. Al-Jaroodi, J., & Mohamed, N. (2019). Blockchain in industries: A survey. *IEEE Access, 7*, 36500–36515.

Chapter 15
Blockchain Technology Acceptance in Agribusiness Industry

C. Ganeshkumar, Arokiaraj David, and Jeganthan Gomathi Sankar

1 Introduction

It has been 10 years (decade) since the pseudonymous author published "Bitcoin: A Peer-to-Peer Electronic Cash Network' white paper." This study laid the groundwork for the rise of Bitcoins, the first cryptocurrency to enable secure financial transactions without needing a dependable central authority, such as financial institutions [36, 38]. With the invention of blockchain technology, Bitcoin addressed the double problem (i.e., the weakness associated with digital tokens since they can be easily duplicated or forged as electronic files).

A blockchain's key advantage is maintaining a clear view and consensus among individuals, although some might not be truthful. Researchers have researched the topic of agreement extensively in the past. Still, its usage in the blockchain context is provided new stimuli, and inspiration leads to new ideas for blockchain systems. The most well-known approach, Proof-Of-work (PoW), is used in Bitcoin. It requires system operators, termed miners, to complete complex computational jobs before confirming transactions and linking them to the Bitcoin blockchain [22, 37]. As a result of this expansion, hundreds of other digital tokens have emerged aimed at addressing some particular vulnerabilities of the dominant cryptocurrencies or targeting certain domains like health, banking, farming, and others. Blockchain is also being studied by the traditional banking system (and in some cases adopted); almost this technology is used for transactions by 15% of financial firms.

C. Ganeshkumar
Indian Institute of Foreign Trade (IIFT), New Delhi, India

A. David (✉)
Al Tareeqah Management Studies, Ras Al Khaimah, UAE

J. G. Sankar
BSSS-Institute of Advanced Studies, Bhopal, India

S. M. Idrees, M. Nowostawski (eds.), *Blockchain Transformations*, Signals and Communication Technology, https://doi.org/10.1007/978-3-031-49593-9_15

Since 2014, it has been more clear that blockchain is used for much more than cryptocurrency and banking transactions enabling the discovery of many new applications: managing and maintaining administrative documents, digital signature, and authentication systems; checking and tracking intellectual property rights and ownership of patent systems; and allowing smart processing. The distributors of locally manufactured goods to the ultimate customer generally control products as they pass from the manufacturer and distributor via the supply chain. These developments are now revolutionizing several areas of industry, society, and government at large, but they can also raise more problems and threads, which need to be expected.

1.1 ICT to the Blockchain

Blockchain is seen as "a transparent, distributed ledger able to effectively and verifiably and permanently record transactions between two parties." Blockchain appears to be a revolutionary ICT with the potential to revolutionize the transforming business utilized in agriculture. On the other hand, farming is an unexplored industry with the potential to transform blockchain completely.

1.2 Blockchain in Agriculture and Food Supply Chain

Although BCT is gaining momentum and proving its usefulness in several cryptocurrencies, different organizations plan to utilize its accountability and error detection to tackle challenges in cases involving several untrusted parties in sharing resources. Food supply chain and agriculture are two significant and very specific fields. Agriculture and food supply chain operations are inextricably linked since agricultural items are frequently utilized as inputs in multi-actor dispersed supply networks, whereas consumers are generally end users. As a positive example, Agri Digital, in December 2016, signed the world's first blockchain sale deal for 23.46 tonnes of grain. About 1300 consumers over 1.6 million tons of grain have since been transacted over the cloud-based network, with $360 million in payments to farmers.

Agri Digital's success encouraged future application within the agricultural supply chain. Agri Digital aims to create reliable and secure supply chains for agriculture through blockchain technology [9, 17]. Worldwide the food supply chain is multi-actors-based and distributed, involving several factors such as producers, logistics firms, retailers, and grocery stores. This system is inefficient and unstable at the moment. For example, when people purchase goods locally, they don't know the source of those products or the production's environmental footprint. Using data and knowledge is becoming important in enhancing production and sustainability for the agricultural sector. Information and communication technology (ICT) improves the efficiency and efficacy of data collection, processing, assessment, and usage in agriculture. It allows farmers to obtain up-to-date information easily and helps them make better daily decisions. For example, data on remotely sensed soil conditions can help farmers manage crops, mobile phones reduce the cost of information and promote farmers' market access and financial support, and introducing a Global Positioning System (GPS) facilitates field mapping and guidance on machinery and crop scouting.

1.3 Crop and Food Production

Address the requirements of a growing population by increasing food production with fewer resources while lowering environmental impact, boosting customer loyalty, preserving responsibility across the supply chain, and ensuring fair pay for farmers despite weather fluctuations, blockchain and IoT are transforming the food processing business from farmer to producer and retailer [2, 16].

Addressing the needs of a growing population by increasing agricultural output despite resource constraints, minimizing environmental imprint, strengthening customer loyalty, maintaining supply chain accountability, and assuring fair remuneration for farmers while managing weather-complexities when it comes to boosting output under adverse environmental circumstances, agriculture has several challenges. Blockchain and IoT are revolutionizing the food manufacturing business from farmer to producer and retailer. Blockchain technology has the potential to make agriculture a more sustainable industry by utilizing a more simplified strategy to optimize farming services such as water, labor, and fertilizer. Here is a detailed explanation of how blockchain technology might help crops or agricultural goods grow.

1.3.1 Step 1: Generating Data from the IoT Devices

Due to the global population being predicted to reach 9.6 billion by 2050, the agriculture business is adding IoT technologies and sensors to feed the expanding population. A sensor network (temperature, pH, soil humidity, atmospheric humidity, and light) designed to monitor the agricultural field. IoT sensors and gadgets capture data that can assist farmers in making crop production decisions. Before saving data, first of all, the collected data from IoT devices has to be analyzed.

1.3.2 Step 2: Enrichment and Cleaning of the Collected Data

There is a need to ensure that the collected data is organized and understandable before saving on the blockchain. Information enrichment is achieved to create greater value and increase the efficiency of collected information. The following steps are taken to ensure that data is cleaned before being distributed to the storage platform:

- **Adding the Meta Information**
- The information of timestamps, demographics, and sorting to make the data more structured.
- **Making the data ready for compliance**
- Preserving blockchain data does not imply that it should not be followed. Consequently, regulatory execution is streamlined. Meeting requirements that guarantee the personally identifiable data associated with IoT system information is protected and complies with safety protocols. It is put into a ready-to-learn format until the data is enhanced.

1.3.3 Step 3: Using Machine Learning (ML) Techniques to Make Data more Insightful

ML is applied for useful insights into the data produced from the sensors. Statistical models will drive the various useful cases including recommendations for the crop quality

- Identification of crop.
- Prediction of crop yield and crop demand.
- From the knowledge collected using deep learning algorithms, farmers and other participants may sometimes improve the irrigation system. The informative data will be stored on the blockchain so that agricultural industry leaders such as innovators, service providers, suppliers, farmers, and retailers may acquire it transparently.

1.3.4 Step 4: Data Kept on the Blockchain

The strong-value information collected through deep learning is maintained in IPFS (Interplanetary File System), a computer cluster network that hashed and stored blockchain addresses. Apart from the traditional method for storing sensitive data on a central server with the potential for one single failure point, the data is shared through all network nodes that prevent central authorities from controlling the device. The data collected on the BCT will cause smart contracts to execute according to the standards prescribed therein. Smart contracts allow data sharing between the various stakeholders in the network stored on the blockchain. Knowledge will be available to any consumer in the agricultural sector, and bringing efficiency in crop or food production will become seamless to them.

1.4 *Food Supply Chain*

Determining the origins or development of food requires tracking the food chain. Supplied eatables must be safe to consume. When it comes to how the food supply chain is really managed, distributors and food producers struggle to confirm the provenance of the commodities. Despite ongoing risks of adulteration, UK households bear an annual cost of up to £1.17 billion due to food fraud. The integration of blockchain technology is introduced to combat this issue, it is now feasible to add transparency and trust to the food supply chain system, assuring food safety for everybody [35]. Here is a detailed description of how the blockchain food supply chain might eliminate dishonesty.

1.4.1 Step 1: Data Generating from IoT Sensors or Farmers Storing Data

As indicated above, smart farming allows sensors to produce relevant information on crops grown in fields. Unless the farmer does not employ technology means, they may readily preserve crucial details such as crop performance, seed types, and climate circumstances under which the crops were cultivated using a cell phone. IPFS with addresses recorded in the blockchain preserves data gathered manually, either by utilizing IoT sensors or by farmers, in a distributed storage network.

1.4.2 Step 2: Distribution of Crop Produce to the Food Processing Companies

Once the harvests are ready, food processing firms will begin purchasing on the site. The commodities may be transported to refineries using IoT-enabled trucks while recording the temperature conditions under which the goods are kept and delivered. If smart contracts approve the bid, the crops are processed, and the data acquired on the blockchain is saved at each stage of the process firms. The information obtained from the refineries will help dealers or distributors determine whether or not the food supplied is of good quality. Saving knowledge about the blockchain will also ensure that protection at any level of the food supply chain has been achieved.

1.4.3 Step 3: Supplying Processed Food to Wholesalers and Retailers

Once the food items or produce are processed, the bidding platform helps wholesalers and distributors to bid on the products they want. The food products are also transported in IoT-enabled vehicles to distributors and supermarkets, equivalent to moving crops to refineries. The blockchain supply chain provides traceability by assisting food manufacturers in simply and effectively carrying out product recalls.

1.4.4 Step 4: Consumers Can Back-Trace the Supply Chain

From information on farm origin to transporting information, sample numbers, food production and manufacturing data, expiry details, inventory weather, and other details automatically connected to food products inside the blockchain, consumers may discover everything by following the supply chain backward. It will enable many stakeholders to have access to food quality information at all levels. Because blockchain increases accountability in the food supply chain, it will be easier to determine when and how food is tainted.

1.5 Controlling the Weather Crisis

Farmers typically face volatile weather conditions when growing various crop types. The prediction and monitoring of climate conditions are therefore important for the survival of crops. Many crops planted in the United States cannot survive floods due to severe spring rainfall. The oxygen concentration hits zero, making life-sustaining activities such as water uptake, root development, and breathing impossible for plants. Furthermore, the existing lack of transparency in the food chain will lead to unpredictable and expensive price surges. Consumers had no notion when crops were harmed by bad weather and what factors contributed to increased pricing. Producers and other stakeholders would better understand pricing disparities in the food delivery sector due to blockchain's capacity to enable monitoring and accountability. Farmers may easily obtain crop insurance premiums using smart contracts if authorized institutions can track climatic conditions from the blockchain ledger.

1.6 A Step-by-Step Blockchain Process for Weather Control in Agricultural Fields

1.6.1 Step 1: Agricultural Climate Stations Submit Necessary Data to the Blockchain

Smart agriculture helps farmers understand the crop's behavior through sensor deployment and field mapping. Placing agricultural climate units on farms may generate vital information such as rainfall, soil temperature, temperature at the dew point, wind speed, air temperature, solar radiation, wind speed, humidity, leaf wetness, direction, and atmospheric pressure at various heights. The abovementioned criteria are calculated, recorded, and maintained in a database that farmers and other authorized entities can easily access.

1.6.2 Step 2: Farmers Can Take the Preventive Measures

Farmers may make educated farm-related decisions by analyzing weather station data. For example, knowing that it will rain hard in the next few days may enable them to take the necessary precautions ahead of time.

1.6.3 Step 3: Crop Insurance for Rapid Application

Farmers may use the blockchain to apply for the rest of their crop insurance claims during a weather disaster. Because of the blockchain's explicit and irreversible activities, insurance companies and other permitted organizations would have simple access to smart weather station data produced. With the aid of smart contracts, it can ask the blockchain directly to obtain the necessary information. After accepting the insurance claim request, farmers will immediately receive the required amount in their wallets. Hence, a blockchain-enabled system can help producers get compensation easily and rapidly.

1.7 Managing the Agricultural Finance

Some of the numerous issues that threaten organized financial inclusion with small farmers are the lack of transparency, payment history, and contract enforcement problems. Exempting financial inclusion will negatively impact agricultural value chain production because producers cannot maximize yields, and purchasers struggle to perform and enhance product supply. Financial firms not only allow smallholders to invest in agriculture but also help farmers to ease liquidity pressures. It resulted in purchasers having difficulties paying farmers on arrival, forcing small farmers to sell their produce at cheaper prices. With transparency and accessibility of shared controls, blockchain brings equity to the agricultural finance process. Here's how blockchain could be able to handle farm finances better.

1.7.1 Step 1: Stakeholders Share the Information at Each Stage of Food Production

Information is saved on the blockchain when a transaction occurs, allowing all interested parties to examine each transaction transparently. The interchange of relevant knowledge would improve the overall system's justice at any level of food production.

1.7.2 Step 2: Auditors Can Conduct Audits Effectively

The blockchain, which can store data indefinitely and safely, can also serve to verify the recorded transactions. Instead, then making producers or retailers call for annual records for audit purposes, auditors can examine payments immediately using database ledgers. The automated auditing approach will render the audit environment at a low cost. Audit firms that can undertake audits throughout the year rather than only at the end of the year. Blockchain allows auditors to replace random auditing, making access to every transaction more effective.

1.8 Benefits of Blockchain Technology in Agriculture

It enables peer-to-peer transactions to be done fairly without needing an intermediate such as a bank (such as cryptocurrency) or an agricultural intermediary. By eliminating the necessity for a centralized authority, technology transforms religion's direction – confidence is placed in encryption and peer-to-peer networks rather than an authority. It helps to restore trust between consumers and producers, which can reduce the cost of payments in the agri-food market. Blockchain platform provides a secure method for tracing anonymous participants' transactions. It allows for the quick identification of fraud and malfunctions. In addition, integrating smart contracts will report problems in real time. Owing to the complexities of the agri-food network, it enables tackling the problem and monitoring of goods in the large supply chain. As a result, technology answers food quality and health issues that are of major interest to consumers, governments, and others.

Blockchain technology ensures responsibility for all parties involved and allows precise data collection. Blockchain technology can track every step in a product's value chain, from conception to demise. Precise agricultural process data is crucial for developing smarter and more secure farming. It enables the implementation of data-driven facilities and compensation solutions in agriculture. Advantages of trying out the blockchain in agriculture are as follows.

1.8.1 Improved Food Quality Control and Safety

One of the primary applications of blockchain is to improve supply chain accountability. It will assist us in eliminating unnecessary procedures and ensuring optimal quality control conditions. Crop failure, for example, is a common issue that affects farmers all over the world. It is mainly due to unfavorable climatic circumstances such as uneven rainfall and variable weather patterns. To address this, companies like IBM are investing millions in precision farming, developing IoT systems that allow producers to assess soil fertility, insects, and water management that may impact their harvests. It will enable them to verify that all variables are as they should be. When something goes wrong, they are quickly alerted and can make corrections until it is too late. Most significantly, it also allows us to identify the cause of the issue almost immediately in the event of the problems such as an outbreak of food safety. It has the potential to save the supply chain cycle more than just time and money – it might also save lives.

1.8.2 Increase of Traceability in the Supply Chain

We are seeing a drastic increase in customer perceptions about food quality. Many customers eagerly wanted to know where their food came from. Customers can know exactly where their food originated from, who cultivated it, and how new it is by utilizing blockchain. Updating the database with details will only allow staff to search for the product at each process point. Consider buying something from the grocery and finding out precisely what it is by scanning a barcode with your smartphone app. Increasing the supply chain's traceability would greatly eliminate food theft, false labeling, and taking out intermediaries from the process to ensure that farmers are paid reasonably for their efforts and that the consumers know what they produce. Improved traceability may also allow farmers to record, monitor, and track their crop status through the growing, harvesting, storing, and delivery process simply by using a smartphone app. They can still see the exact state of their goods and make changes where appropriate. Realizing that over one-third of food generated for human use is wasted each year (roughly 1.3 billion tonnes), this innovation is long overdue. Some Australian farmers are adopting blockchain to track their produce and reduce waste.

1.8.3 Increase in Efficiency of the Farmers

To save information and manage procedures, many farmers now utilize a range of programs, spreadsheets, and notes produced by various software development firms. However, sending this information to other network providers is complex and time-consuming [41]. Farmers can use blockchain technology to keep information in a single spot so that people who need it can access it, automate the entire process, and save significant time and energy. For example, they might monitor the things like:

- The business objectives and how they intend to accomplish those goals
- Animal health issues, what they eat, and how much they should be fed
- Income and expenses
- Their schedule, how much each employee must be compensated, and also how many hours each employer must work

Keeping track of everything in one application rather than a variety of ways simplifies the operation and avoids the chance of losing vital data.

1.8.4 Fairer Payment for the Farmers

Several obstacles make it hard for farmers to be compensated for their produce. For starters, it takes farmers several weeks to get full Payment for their produce. To make matters worse, traditional payment channels – particularly wire transfers – consume a sizable chunk of farmers' revenue. Blockchain-based smart contracts automatically trigger payments when a buyer meets certain criteria without charging high transaction costs. It suggests that the farmer may receive money for the items as soon as they are sent without losing significant revenues. Many farmers sometimes struggle to sell their produce at a respectable price on the market. The intermediaries are now reaping most benefits while doing little labor. Smart contracts will eliminate the need for intermediaries, allowing farmers to speak directly with distributors. As a result, they may obtain a more equitable price for their goods.

1.9 Applications

Various uses in food and agriculture: farm insurance, digital farming, food supply chain, and agricultural product purchasing.

1.10 Agriculture Insurance

Extreme climate affects the production of agriculture and endangers food security. Both livestock and crop production get involved and change in temperature is predicted to intensify more in future weather extremes. Agricultural insurance policies are commonly recognized as an instrument for managing climate-related risks. Before the cropping process starts, farmers pay an insurance premium here and earn an insurance payout if they suffer a loss on their farm. Consequently, the insurer absorbs all the liability insurance liability, and farmers can control the economic exposure to severe weather events and financial losses incurred in extreme climatic conditions. In fact, in the event of weather concerns that consistently affect all covered farms, the insurer will further hedge the systemic element of the risk with a reinsurance provider. Agricultural insurance varies in the calculation of risks and, subsequently, in the initiation of payouts. Insurances that reward farmers based on a damage estimate rendered by a farm specialist are referred to as compensation-based insurance. Indemnity-based insurance can cover losses specifically but is vulnerable to asymmetric information problems. More precisely, information about the risk of agricultural production and business practices is transmitted asymmetrically between farmers and insurers. Farmers should be taught about both moral hazard considerations and unfavorable choices. According to the adverse selection, farmers with a high-risk exposure are more likely to get insurance than farmers with a lower-risk exposure. When protected, farmers are more likely to engage in riskier production strategies. Both conditions contribute to insurance policy business loss if the insurer lacks knowledge. Therefore, insurance dependent on reimbursement is vulnerable to expensive risk assessment, and steps need to be introduced to prevent asymmetric knowledge issues, such as deductibles. In addition, outputs cannot be assessed, for example, weathered pastures cannot be covered, even though they cause financial harm.

Index-based insurance was born, inspired by the disadvantages of compensation-based insurance, either as an alternative or as a supplement to classic products. In this case, the payout is driven by a visible metric, such as rainfall at a local weather station, rather than the loss itself. If this weather station has adequate prior weather records, the farmer and the insurer have identical insured interest details, and agricultural operations do not influence insurance payouts.

Consequently, moral hazard and adverse selection have no part to play, and the technological process for inducing a payout has been simplified considerably. In addition, full insurance coverage is available without deductibles, and claims can be made promptly and automatically only after an extreme weather occurrence is assessed. The risk basis is the difference between payout and on-farm loss. There may be three Specific Risk Sources. Spatial risk bases suggest variations between on-farm and measured conditions, for example, spatial distance. Temporary risk means that an imprecise period has been chosen to determine the index, for example, year-round rainfall vs. Seasonal rainfall in May. All remaining factors, such as a lack of meteorological variables or biased technical implementation, are summarized in the design foundation risk [47].

To sum up, index insurance is becoming an increasingly valuable method for farmers to mitigate risk, although raising risk on the basis is of central interest. Blockchain can aid in the two-dimensional improvement of index insurance. First, payments can be made based on weather data, which triggers payout as described in a smart contract timely and automatic. Second, environmental and other information sources, such as plant growth data or data acquired by farm machinery, may be automatically merged into a smart oracle, which enhances risk management through indexing and payout procedures. Smart contracts employing digital oracles to combine external data have proven beneficial in various crypto-economic applications.

1.11 Smart Agriculture

Critical data and natural resource information are underlying the agri-food systems that support all farming forms. Stakeholders and actors produce and handle data and knowledge according to their capacities and needs. Smart farming is differentiated using ICT, the IoT, and many developing data collecting and analysis technologies like machine learning sensors and UAVs. An important problem in developing smart farming is implementing a robust protection framework that promotes data utilization and management. Modern forms of centralized information processing are vulnerable to misleading data, information manipulation and misuse, and cyber-attacking. For example, government-centralized agencies usually handle environmental monitoring data with their interests. They can exploit data-related decision taking. It stores data generated by many actors and stakeholders throughout the value-added cycle of creating an agricultural product, from seed to sale. It guarantees that the data acquired is permanent and accessible to all actors and stakeholders. Blockchain technology generates protection through the "defence of anonymity" rather than decentralization on which conventional technologies depend. It is less vulnerable to loss of data and manipulation to send information to stakeholders' computers than to store data on servers centrally controlled by administrators.

The adoption of IoT and blockchain technology increased recently, which has led to creating and deploying various smart farming agriculture models. Patil, for example, suggests "a lightweight blockchain-based architecture for smart greenhouse farms." ICT e-farming concept based on blockchain for local and regional use has been proposed by Lin et al. [26]. Several businesses are dedicating themselves to smart agriculture blockchain technology. Filament manufactures equipment that connects physical items with networks using smart farming technology. Transparency inspires the public's commitment to irrigation management and strengthens their work to advance water resource utilization. Over time, the quantitative database created using blockchain may be employed to assist decision-making regarding the water canals' construction and maintenance.

Smart blockchain farming doesn't lower the technical barriers to the farmers to participate, if not lift them. Importantly, gathering trustworthy data from big farmers is more encouraged than from smallholders to upload it to the technology of blockchain. On a larger scale, farmers often encounter challenges that can affect their profits in smart farming using blockchain technology. It can thus establish or intensify the difference between big and small farmers.

1.12 Food Supply Chain

Food supply networks got more complicated and lengthened as globalization and market demand increased. There are a few unique difficulties in food supply chains, such as food traceability, safety and quality, customer trust, and supply chain inefficiencies, which adds extra dangers to society, the economy, and human health.

From the standpoint of producers, the use of blockchain technology will assist in developing confidence with customers and the legitimacy of the items by giving explicit information about specific products in blockchain. Enterprises can better attain the value of their goods and thereby improve productivity. It will make it impossible for fraudulent suppliers and less-quality goods to remain in the markets and push all the suppliers to increase the quality of the goods in the agriculture and food sectors. It contributes to the availability of precise and reliable information on how food is prepared and exchanged from the client's perspective. It assists in resolving customer issues about food safety, quality, and environmental friendliness. Blockchain allows customers to interact with farmers since they can better understand the food production cycle. It helps customers by removing barriers to product interchange to strengthen their connection, ultimately enhancing customer confidence and faith in food health. From the standpoint of regulatory authorities, blockchain supplies them with secure and correct information to adopt intelligent, effective policies.

Blockchain can record a product's data from its origin to retail location. It provides a secure and consistent method of storing information collected at the start of the supply chain, such as grain or vegetable pesticide residues and livestock animal DNA. Individuals involved in the product's supply chain can confirm and review these details. Obtaining this data for all objects might be expensive, but it can be done on samples. The availability of such information will aid in detection.

In the food supply chain, the latest blockchain innovation is still in its beginning stages of advancement. Simultaneously, the methodology of executing blockchain innovation contains various holes and incomplete business [29]. Moreover, the arrangement of blockchain innovation requires broad support and coordinated effort of partners in the food chain, which is basic for its motivation. Considering its straightforwardness, security, and decentralization attributes, blockchain innovation considers observing the nature of food and data across the whole supply chain. It dissuades food buy extortion and cuts the cost of running the food supply chain. Accordingly, all gatherings, including makers, clients, and government administrative associations, will benefit.

1.13 E-Commerce of Agricultural Products

Agricultural commodities and trade e-commerce are confronting some significant challenges. First, as Tiago et al. [43] showed, highly confident customers are more inclined to purchase online. Consumers cannot easily confirm and trust the basic details about agricultural products [20]. Meanwhile, the most important challenges e-commerce companies facing, particularly in developing countries, are logistics service and cash on delivery. In addition, retailers in e-commerce often need to manage time-consuming small orders of different products, which creates high operating costs for e-commerce firms.

Blockchain technology has the potential to solve several of these problems: (1) Data security – blockchain enables essential validation and authentication needs as a powerful instrument. Thus, it can safely and unchangeably link data on all aspects of the farming and harvesting of agricultural produce. (2) Supply chain management. Blockchain technology will more effectively allow supply chain management than conventional control methods by reducing the signaling cost for each entity. Any link in the supply chain – the supplier, the point of origin, the logistics business, the location, multi-modal transportation, the container, and the last mile – represents a "block" of information with the benefits of visibility, aggregation, confirmation, automation, and resilience. (3) Methods of Payment – the blockchain offers a zero-rate, encrypted solution to payment. In addition, using cryptocurrencies to sell agricultural goods will more dramatically lead to reduced transaction costs. (4) Customer trust. The distributed accounting information system of the blockchain is timestamped during the decentralized process, ensuring that all information on the chain is visible and unalterable. Consumers will be free of imposters, and they will re-establish faith in e-commerce. (5) Farmers' costs are reduced. Households create numerous agricultural commodities; yet, due to the low quantity and small scale of operations, traditional e-commerce is unwilling or unable to provide services to them, thus eliminating such players from the trade [13, 40]. Blockchain is a technology to significantly lower transaction costs and reintegrate them into the market.

Few businesses also use this technology for testing, but it may not be used in the entire process. Detailed details, including seeding, irrigation, fertilization, and deworming, were documented before the products were placed into the network. We also have basic producer information, transport logistics, day storage, and temperature storage. Only the special QR code on the items can be scanned by customers, and all the details are available for search. This strategy will successfully halt counterfeiting by rogue merchants and restore customer trust in e-commerce and agricultural products from its suppliers.

2 Review of Literature

Ammous et al. [3] explain that given the complexity of current supply chains, BT must automate and improve process efficiency. There are some facilities in a typical supply chain, such as distributors, factories, distribution centers, etc., through which flows products, cash, and information [45]. At these interfaces, few transactions take place in all supply chains, for example, between a salesman and a plant or between a factory and a distribution center [9]. In the presence of financial-related redundancies in record keeping, trust-related problems are inevitable.

Korpela et al. [23] state that the blockchain will serve as the primary source of information and will encompass all supply chain operations. Many value-creating processes, like capture, monitoring, and data transfer, may be accelerated, and blockchain can provide scalability without delay [42]. In their supply chain network, companies will get a distributed real-time log of transactions and operations through a blockchain-powered supply chain.

Ivanov et al. [19] propose that BT helps companies obtain accurate forecasts of demand, manage capital effectively, and reduce the cost of carrying inventories due to their ability to track operations. It allows supply chains to reduce cost risk compared to conventional supply chains where high inventory levels, overcapacity, and third-party backup sources are built in anticipation of interruptions.

According to Sreehari et al.'s [39] report, the blockchain can introduce digital confidence into the procurement cycle. Blockchain's transparency characteristic comes in handy here. In conventional supply chains led by analog contracts, there is a payment gap between the actual delivery of the goods, the creation of invoices, and the ultimate settlement of payments. BT's smart contracts deployment will help businesses decrease or eliminate this delayed and costly payment gap by integrating delivery and payment into enterprise-wide digital contracts and connecting with logistics partners and banks [34]. The integration will also reduce the need for working capital and simplify the financing operations that contribute to supply chain sustainability. A smart contract, which serves as a rule book for transactions, can be employed. These intelligent agreements can aid payment decisions [27]. Vehicles, for example, which can be readily integrated with technology such as GPS, can serve as a blockchain information input source. After such inclusion, the irreversible presence of blockchain ensures that it does not change data and may be utilized for analytics purposes at any time. The combination boosts operating efficiency for the outbound supply chain even further.

Kshetri [24] makes sense of how different SCM standards, like moderateness, quality, practicality, steadfastness, risk the executives, maintainability, and versatility, were considered. Utilizing innovation diffusion theory, the review found that organizations, for example, oil exchanging with a few provider layers, will arise as the leaders in blockchain acknowledgment [32]. The review anticipated that exposure of a blockchain by one association would force prescriptive tension on other supply chain elements involving ranchers, for instance.

Nakasumi [28] discusses the approach to addressing conventional supply chain issues like double marginalization and asymmetry of information, etc. The study suggests that the problem of double marginalization and knowledge asymmetry can be solved by applying BT. A cost-effective, scalable, cloud-based integration model is proposed [18]. According to the survey, the most crucial blockchain functions are ledgers, smart contracts, and time stamps.

Tian's [44] study presents a food supply chain traceability system for real-time food monitoring that is based on HACCP (Hazard Analysis and Critical Control Points), blockchain, and the Internet of Things [33]. The suggested approach provides the advantages of monitoring and detecting fake products. Blockchain is believed to be able to revolutionize the food chain. Abeyratne and Monafared's [1] study uses a cardboard box as an example to highlight the diverse advantages of blockchain in different industrial application domains within the supply chain. According to the analysis, IoT and blockchain will significantly influence manufacturing in the coming decade [8, 25].

Kamble et al.'s [21] study delves into the comprehensive exploration of blockchain technology and its significance throughout the supply chain. They establish and validate a model for assessing customer expectations regarding blockchain adoption. The idea depends on the assembly of three reception theories: the technology acceptance model (TAM), the technological readiness index (TRI), and the anticipated conduct theory (TPB). A review of 181 supply chain specialists in India was utilized to test the recommended model utilizing underlying condition modeling. The study found that the TRI-Insecurity and inconvenience affected apparent ease of use or benefit. The review indicates that the TRI develops uncertainty and uneasiness, significantly affecting apparent ease of use and benefit. The motivation to take action is influenced by perceived benefits, mindset, and perceived control over activities.

The report Astarita et al. [5] offers the consequences of a writing survey using blockchain-based transportation frameworks. The significant objective was to utilize a multi-step cycle to distinguish ebb and flow research patterns and possible future troubles. Following that, a bibliometric survey was finished to acquire a wide understanding of the subject of interest. Following that, the main headways were completely analyzed, zeroing in on two regions: supply chain and planned operations, street traffic, the executives, and brilliant city, the board. The essential outcome is that blockchain technology is still in its beginning stages. Yet it seems very encouraging, considering its possible applications in spaces, for example, food checking and recognizability, administrative consistence, shrewd vehicle security, and supply-demand balance.

Cole et al. [7] explain the review of blockchain technology from the Operations and Supply Chain Management (OSCM) perspective, finding application areas. A description and analysis of the blockchain technology are provided to identify implications for the OSCM field. There is strong excitement about the possibilities provided by digital ledger technology [4]. The nascent state of practice and research surrounding blockchain means that OSCM researchers can study and influence the acceptance of the technology at its early stages. Today, there's more scope for creativity now than ever.

2.1 *Objectives of the Study*

1. To understand the perceived ease of using the blockchain
2. To study the relationship between the components of the blockchain
3. To understand the perceived risk of using the blockchain
4. To examine the intention to use blockchain

2.2 *Hypotheses of the Study*

H_1: There is a significant association between perceived risks and intention to use.

H_2: There is a significant association between perceived ease of use and perceived usefulness.

H_3: There is a significant association between perceived usefulness and intention to use.

H_4: There is a significant association between perceived ease of use and intention to use.

3 Research Methodology

The present study was undertaken in Bangalore city of Karnataka, with 50 responses from the employees of Agribusiness companies. The selection of the sample is based on a multi-stage sampling technique. In the first stage, companies were selected purposively related to agriculture in Bangalore. In the second stage, departments in the companies were set purposively associated with blockchain. In the third stage, the employees are chosen randomly.

A structured questionnaire was developed to collect primary data from the employees of companies. The data collected from each respondent about know the perceived usefulness of blockchain, intention to use blockchain, perceived risk of blockchain, and perceived ease of use of blockchain [10–12, 14, 18]. Based on information collected from the respondents, technology acceptance model is to be assessed. Secondary data has also been collected from various online sources related to the blockchain. The data collected from the surveyed area in Bangalore has been analyzed using descriptive statistics and the PLS-SEM model.

The descriptive statistics of the gender were categorized into male and female. More than half of the respondents (62%) are male and 38% are female. Nearly half of the respondents are (48%) postgraduate degree equivalent, 28% are undergraduate, (20%) respondents are others, and the remaining 4% have completed their doctorate. Of most respondents, 78% had heard about blockchain, and the majority, 84%, stated that blockchain would be beneficial. The majority of respondents, 82%, marked that blockchain technology is helpful for future purposes, and half of the respondents (50%) stated that it helps them to complete their tasks quickly. Almost 44% are marked neutral about using blockchain to improve job performance, and 8% strongly agree that blockchain enhances job performance, while 6% of the respondents disagree.

Almost half of the respondents (50%) agree that blockchain improves job productivity, and most (42%) agree that using blockchain-based applications increases effectiveness on the job. More than half of the respondents (54%) agreed that using BBA makes it easier to do their job. Most respondents (74%) have stated that neither easy nor difficult to learn how to operate blockchain. Less than half of the respondents (44%) have said that neither easy nor difficult to get BBA in what they want to do. Most of the respondents (60%) are marked neutral to the interactions with the BBA would be clear and understandable, and 58% of them agree that BBA is to be flexible to interact with.

More than half of the respondents, 54%, agreed that it would be easy to become skillful using BBA, and 58% agreed they worry about transaction errors and cannot get compensation loss. Most of them (72%) have marked neutral with BBA may not perform well because of technical errors in the network. Most (68%) are neutral to the blockchain, may not perform well, and process payments incorrectly.

More than one-third (32%) of the respondents agree that they would not feel safe providing personal or private information over blockchain-based applications, and 48% of them have marked neutral to the usage of blockchain, leading to loss of convenience due to wasting time fixing errors. The majority of them (70%) have observed that neutral to learning how to use blockchain takes a lot of time, and 68% of them are desirable to learn how to use blockchain. Almost 48% are neutral about using blockchain for banking needs, and 56% are neutral about using blockchain for transactions. More than half of the respondents (52%) agree they feel comfortable using blockchain for transactions (Fig. 15.1 and Table 15.1).

The table above explains the relationship between the perceived risk to intention to use, perceived ease of use to perceived usefulness, perceived usefulness to intention to use, and perceived ease of use to intention to use is positive at 0.05 Significance level.

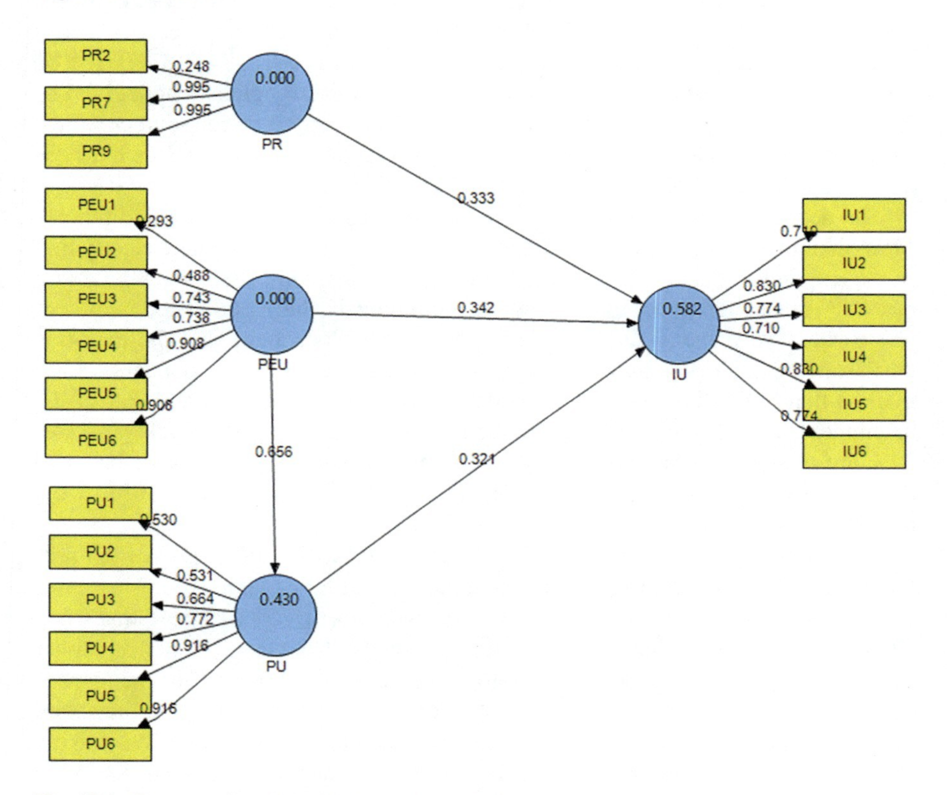

Fig. 15.1 Conceptual model with beta value statistics

Table 15.1 Results of hypothesis testing

	Relationships	T-value	Beta value	Hypothesis results
H_1	Perceived risk to intention to use	3.34	0.33***	Supported
H_2	Perceived ease of use to perceived usefulness	9.70	0.65***	Supported
H_3	Perceived usefulness to intention to use	2.80	0.32***	Supported
H_4	Perceived ease of use to intention to use	3.46	0.34***	Supported

Note *** highly significant

4 Discussion and Conclusion

Most respondents from the surveyed area are male at the postgraduate level. Many respondents have heard of blockchain technology and its applications and benefits. There will be a usage of blockchain of technology in the future, which helps to ensure traceability. Half of the respondents stated that the blockchain-based applications help to complete the tasks quickly, and some said there is no change in completing the tasks. The majority of the respondents from the surveyed explained that there is no change in their job performance by using the blockchain. By using blockchain-based applications, there is an increase in productivity and effectiveness on the job, and it helps to complete the job easily. Most respondents worry they cannot get compensation when a transaction error occurs. Over one-fourth of respondents from the surveyed area are not interested in providing personal information in blockchain-based applications. Most respondents desire to learn blockchain-based applications and are neutral to using blockchain for banking needs. The components of blockchain have a favorable connection, including perceived ease of use, perceived utility, intention to use, and perceived danger [6, 31, 46]. A large percentage of responders use blockchain to handle transactions.

It concluded that blockchain technology acceptance had been seen. blockchain helps to ensure traceability and also helps to increase productivity. blockchain is neither easy nor difficult to learn its applications. By using blockchain, they may become skillful and interested in providing personal information. They are neutral that blockchain may not perform well because of technical errors and for banking needs. It is used for handling transactions. There is a positive relationship between the components of blockchain, such as intention to use, perceived use, perceived usefulness, and perceived risk [15, 30]. The study focuses on the technology acceptance model of blockchain in various agriculture-related companies. A survey has been carried out in the companies that are located in Bangalore, such as Cropin, Smerketo, Spar, Reliance, Big Basket, Jivabhumi, Jubilant food works, Way Cool, Licious, Cultvyate, GT green technologies, Farmtrac, Sammunathi, etc.

References

1. Abeyratne, S. A., & Monfared, R. P. (2016). Blockchain-ready manufacturing supply chain using a distributed ledger. *International Journal of Research in Engineering and Technology, 5*(9), 1–10.
2. Ahmed, I., Mehta, S. S., Ganeshkumar, C., & Natarajan, V. (2022). Learning from failure to enhance performance: A systematic literature review of retail loss. *Benchmarking: An International Journal*, (ahead-of-print).
3. Ammous, D., Chabbouh, A., Edhib, A., Chaari, A., Kammoun, F., & Masmoudi, N. (2023). Improved YOLOv3-tiny for silhouette detection using regularisation techniques. *International Arab Journal of Information Technology 20*(2), 270–281.
4. Arokiaraj, D., Ganeshkumar, C., & Paul, P. V. (2020). Innovative management system for environmental sustainability practices among Indian auto-component manufacturers. *International Journal of Business Innovation and Research, 23*(2), 168–182.

5. Astarita, V., Giofrè, V. P., Mirabelli, G., & Solina, V. (2020). A review of Blockchain-based systems in transportation. *Information, 11*(1), 21.
6. Biswas, M., Akhund, T. M. N. U., Ferdous, M. J., Kar, S., Anis, A., & Shanto, S. A. (2021, July). Biot: Blockchain-based smart agriculture with internet of thing. In *2021 fifth world conference on smart trends in systems security and sustainability (WorldS4)* (pp. 75–80). IEEE.
7. Cole, R., Stevenson, M., & Aitken, J. (2019). Blockchain technology: Implications for operations and supply chain management. *Supply Chain Management: An International Journal, 24*(4), 469–483.
8. David, A. (2020). Consumer purchasing process of organic food product: An empirical analysis. *Journal of Management System-Quality Access to Success (QAS), 21*(177), 128–132.
9. David, A., Ravi, S., & Reena, R. A. (2018). The eco-driving behaviour: A strategic way to control tailpipe emission. *International journal of. Engineering & Technology, 7*(3.3), 21–25.
10. David, A., Nagarjuna, K., Mohammed, M., & Sundar, J. (2019a). Determinant factors of environmental responsibility for the passenger car users. *International Journal of Innovative Technology and Exploring Engineering (IJITEE), 9*, 210–224.
11. David, A., Thangavel, Y. D., & Sankriti, R. (2019b). Recover, recycle and reuse: An efficient way to reduce the waste. *International Journal of Mechanical and Production Engineering Research and Development, 9*, 31–42.
12. David, A., Kumar, C. G., & Paul, P. V. (2022). Blockchain technology in the food supply chain: Empirical analysis. *International Journal of Information Systems and Supply Chain Management (IJISSCM), 15*(3), 1–12.
13. Ganeshkumar, C., & David, A. (2022, August). Digital information Management in Agriculture—Empirical Analysis. In *Proceedings of the third international conference on information management and machine intelligence: ICIMMI 2021* (pp. 243–249). Springer Nature Singapore.
14. Ganeshkumar, C., Prabhu, M., & Abdullah, N. N. (2019). Business analytics and supply chain performance: Partial least squares-structural equation modeling (PLS-SEM) approach. *International Journal of Management and Business Research, 9*(1), 91–96.
15. Ganeshkumar, C., Prabhu, M., Reddy, P. S., & David, A. (2020). Value chain analysis of Indian edible mushrooms. *International Journal of Technology, 11*(3), 599–607.
16. Ganeshkumar, C., Jena, S. K., Sivakumar, A., & Nambirajan, T. (2021). Artificial intelligence in agricultural value chain: Review and future directions. *Journal of Agribusiness in Developing and Emerging Economies*. https://doi.org/10.1108/JADEE-07-2020-0140
17. Ganeshkumar, C., David, A., & Jebasingh, D. R. (2022). Digital transformation: Artificial intelligence based product benefits and problems of Agritech industry. In *Agri-Food 4.0*. Emerald Publishing Limited.
18. Ganeshkumar, C., David, A., Sankar, J. G., & Saginala, M. (2023). Application of drone Technology in Agriculture: A predictive forecasting of Pest and disease incidence. In *Applying drone technologies and robotics for agricultural sustainability* (pp. 50–81). IGI Global.
19. Ivanov, D., Dolgui, A., & Sokolov, B. (2019). Digital technology and Industry 4.0 impact the ripple effect and supply chain risk analytics. *International Journal of Production Research, 57*(3), 829–846.
20. Jeganathan, G. S., David, A., & Ganesh Kumar, C. (2022). Adaptation of Blockchain technology in HRM. *Korea Review of International Studies, 15*, 10–22.
21. Kamble, S., Gunasekaran, A., & Arha, H. (2019). Understanding the Blockchain technology adoption in supply Chains-Indian context. *International Journal of Production Research, 57*(7), 2009–2033.
22. Kamilaris, A., Fonts, A., & Prenafeta-Boldΰ, F. X. (2019). The rise of blockchain technology in agriculture and food supply chains. *Trends in Food Science & Technology, 91*, 640–652.
23. Korpela, K., Hallikas, J., & Dahlberg, T. (2017, January). Digital supply chain transformation toward blockchain integration. In *Proceedings of the 50th Hawaii international conference on system sciences*.

24. Kshetri, N. (2018). 1 Blockchain's roles in meeting key supply chain management objectives. *International Journal of Information Management, 39*, 80–89.
25. Latha, C. J., Sankriti, R., David, A., & Srivel, R. (2020). IoT based water purification process using ultrasonic aquatic sound waves. In *Test Engineering & Management*. The Mattingley Publishing Co., Inc. ISSN, 0193-4120.
26. Lin, Y. P., Petway, J. R., Anthony, J., Mukhtar, H., Liao, S. W., Chou, C. F., & Ho, Y. F. (2017). Blockchain: The evolutionary next step for ICT e-agriculture. *Environments, 4*(3), 50.
27. Madaan, G., Swapna, H. R., Kumar, A., Singh, A., & David, A. (2021). Enactment of sustainable technovations on healthcare sectors. *Asia Pacific Journal of Health Management, 16*(3), 184–192.
28. Nakasumi, M. (2017, July). Information sharing for supply chain management based on block chain technology. In *2017 IEEE 19th Conference on Business Informatics (CBI)* (Vol. 1, pp. 140–149). IEEE.
29. Nehme, E., El Sibai, R., Bou Abdo, J., Taylor, A. R., & Demerjian, J. (2022). Converged AI, IoT, and blockchain technologies: A conceptual ethics framework. *AI and Ethics, 2*(1), 129–143.
30. Pachayappan, M., Ganeshkumar, C., & Sugundan, N. (2020). Technological implication and its impact in agricultural sector: An IoT Based Collaboration framework. *Procedia Computer Science, 171*, 1166–1173.
31. Panpatte, S., & Ganeshkumar, C. (2021). Artificial intelligence in agriculture sector: Case study of blue river technology. In *Proceedings of the second international conference on information management and machine intelligence* (pp. 147–153). Springer.
32. Poongodi, T., Ilango, S. S., Gupta, V., & Prasad, S. K. (2022). Influence of blockchain technology in pharmaceutical industries. In *Blockchain technology for emerging applications*. https://doi.org/10.1016/B978-0-323-90193-2.00009-0
33. Pratheepkumar, P., Sharmila, J. J., & Arokiaraj, D. (2017). Towards mobile opportunistic in cloud computing. *Indian Journal of Scientific Research (IJSR), 17*(2), 536–540. ISSN: 2250-0138.
34. Ravi, S., David, A., & Imaduddin, M. (2018). Controlling & calibrating vehicle-related issues using RFID technology. *International Journal of Mechanical and Production Engineering Research and Development, 8*(2), 1125–1132.
35. Sadanandam, A., Abraham, S., David, A., & Ramasamy, R. (2022, April). A study on effect of virtual reality on tourist buying behavior-an empirical analysis. In *International conference on "Trends & disruptions in hospitality & tourism" 22nd–24th April 2022, conference proceedings, black eagle books* (pp. 308–318).
36. Sheel, A., & Nath, V. (2019). Effect of blockchain technology adoption on supply chain adaptability, agility, alignment and performance. *Management Research Review, 42*(12), 1353–1374.
37. Siddhartha, T., Nambirajan, T., & Ganeshkumar, C. (2019). Production and retailing of self-help group products. *Global Business and Economics Review, 21*(6), 814–835.
38. Siddhartha, T., Nambirajan, T., & Ganeshkumar, C. (2021). Self-help group (SHG) production methods: Insights from the union territory of Puducherry community. *Journal of Enterprising Communities: People and Places in the Global Economy*. https://doi.org/10.1108/JEC-01-2021-0005
39. Sreehari, P., Nandakishore, M., Krishna, G., Jacob, J., & Shibu, V. S. (2017, July). Smart will converting the legal testament into a smart contract. In 2017 International Conference on Networks & Advances in Computational Technologies (NetACT) (pp. 203-207). IEEE.
40. Sreenu, M., Gupta, N., Jatoth, C., Saad, A., Alharbi, A., & Nkenyereye, L. (2022). Blockchain based secure and reliable Cyber Physical ecosystem for vaccine supply chain. *Computer Communications, 191*, 173–183.
41. Srivastava, V., Singh, A. K., David, A., & Rai, N. (2022). Modelling student employability on an academic basis: A supervised machine learning approach with R. In *Handbook of research on innovative management using AI in industry 5.0* (pp. 179–191). IGI Global.

42. Thummula, E., Yadav, R. K., & David, A. (2019). A cost-effective technique to avoid communication and computation overhead in vehicle insurance database for online record monitoring. *International Journal of Mechanical and Production Engineering Research and Development (IJMPERD), 9*(2), 711–722.
43. Tiago, T., Couto, J. P., Tiago, F., & Faria, S. D. (2017). From comments to hashtags strategies: Enhancing cruise communication in Facebook and Twitter. *tourismos, 12*(3), 19–47.
44. Tian, F. (2017, June). A supply chain traceability system for food safety based on HACCP, Blockchain & Internet of things. In *2017 international conference on service systems and service management* (pp. 1–6). IEEE.
45. Tönnissen, S., & Teuteberg, F. (2020). Analysing the impact of blockchain-technology for operations and supply chain management: An explanatory model drawn from multiple case studies. *International Journal of Information Management, 52*, 101953.
46. Victer Paul, P., Ganeshkumar, C., Dhavachelvan, P., & Baskaran, R. (2020). A novel ODV crossover operator-based genetic algorithms for traveling salesman problem. *Soft Computing, 24*(17), 12855–12885.
47. Xiong, H., Dalhaus, T., Wang, P., & Huang, J. (2020). Blockchain technology for agriculture: Applications and rationale. *Frontiers in Blockchain, 3*, 7.

Chapter 16
Adoption of Block Chain Technology and Circular Economy Practices by SMEs

Mukesh Kondala, Sai Sudhakar Nudurupati, and K. Lubza Nihar

Abbreviations

BC	Blockchain
CE	Circular Economy
GDP	Gross Domestic Product
MSME	Micro, Small, and Medium Enterprises
SDG	Sustainable Development Goals
SME	Small and Medium Enterprises

1 Introduction

The adoption and implementation of digital technologies by organizations have given rise to massive transformations in the organizations' internal operations and processes. The primary objective of Industry 4.0 is to facilitate the seamless integration of technology, electronic devices, transactions, and various stakeholders, including suppliers, consumers, and clients [1]. Blockchain (BC) technology is a state-of-the-art innovation that holds the potential to enhance the reliability and safety of these interconnections [2]. Technology adoption and implementation come with a set of challenges like adverse environmental consequences [3]. The significance of the circular economy (CE) in contributing to the sustainable development goals (SDGs) has been increasing in recent years. According to scholarly works by [4–7]. China has been at the forefront of adopting and implementing the CE model.

The concept of CE centers on enabling diverse processes such as refurbishment, recycling, repair, design, manufacturing, and eco-effectiveness advancement. The overarching objective of these endeavors is to reduce the disposal of materials, components, and commodities. The concept of the CE aims to identify a rationale for

M. Kondala (✉) · S. S. Nudurupati · K. L. Nihar
GITAM School of Business, GITAM (Deemed to be University),
Visakhapatnam, Andhra Pradesh, India

S. M. Idrees, M. Nowostawski (eds.), *Blockchain Transformations*, Signals and Communication Technology, https://doi.org/10.1007/978-3-031-49593-9_16

generating value from waste and resources by transforming consumption and production patterns. This chapter discusses using BC technology in SMEs to adopt and implement a CE.

This chapter is structured as follows: The second section outlines the methodology used. Section 3 identifies gaps in the literature review process. Section 4 summarizes the research findings related to the concept of CE, its significance, challenges, and opportunities for Indian SMEs in adopting CE with BC technologies while also discussing the challenges encountered by small enterprises. Section 5 presents the discussions derived from the comprehensive analysis of the findings and provides recommendations for future research. Section 6 ends the chapter with a conclusion.

2 Methodology

A comprehensive analysis was conducted to clarify the methodology outlined by [8] and identify relevant research pertaining to BC, CE, and sustainability in the context of Indian micro, small, and medium enterprises (MSMEs). Established criteria were applied to select the relevant studies for this systematic literature review. The search term TITLE-ABS-KEY ((“circular economy“AND sustainability AND practices OR sustainability AND performance AND Blockchain OR Technology)) was used as the default search criterion. It is worth noting that there is a scarcity of research on the topics of BC [9] and the circular economy [7, 10, 11]. The findings indicate an increase in the exploration of BC research since 2020, with a growing focus in 2022. While some papers highlight the benefits of adopting CE, there is a lack of scholarly inquiry regarding this approach’s practical applications and resulting impacts. The construction of the literature review followed a rigorous selection process, including only articles that met the specific criteria of the study, ensuring a focused and targeted review.

The authors adopted a systematic literature review (SLR) method, following [12], to gather relevant and precise information for their study. The search and gathering of articles were performed jointly to prevent any mismatches, and the relevant articles were filtered based on the following search criteria.

1. The first search is based on keywords “Circular economy,” “Blockchain,” and “sustainability,” with additional connections mentioned in the search string, which resulted in 297 documents.
2. The second step is based on the application of various filters [13] such as open access journals, limiting the subject area to computers, business, management, and accounting, which retrieved 91 articles.
3. Upnext the articles were further filtered with abstract filtration and criteria for the current chapter such as the paper discussing BC and the CE. Moreover, the papers should contain discussions about the technological perspectives to pro-

mote sustainability and should address the challenges faced by the MSMEs. A total of 54 articles were retrieved by this filter.
4. Further, the cross-references, seminal papers, and methodology papers were identified and added to the final number of articles for the study.

The search is to filter the articles and abstract reading, and the subject area is limited to industrial engineering and management only. The search was performed only in the Scopus database, which is much more relevant than many other databases. The articles were analyzed by the forward and backward snowball technique, which extracted the exact data that fit our study [14].

3 Research Gap

While earlier literature focused on decoupling policies without actual implementation of practices [15], recent research has delved into the complexities of observed implementations [2]. In the context of SMEs in India, there appears to be lack of awareness regarding CE practices and the use of BC technology to reduce waste in production [9]. Industry experts must implement BC technologies to enhance transparency and agility [16]. However, there are still gaps in understanding how SMEs can effectively adopt CE practices [11]. Furthermore, there is a need for more research on the environmental, economic, social, and behavioral effects of circular practices and the adoption of these practices by different stakeholders. By addressing these research gaps, we can accelerate the development and implementation of CE approaches, working toward a more sustainable and circular future [17]. Achieving a technological CE requires collaboration among policymakers, businesses, consumers, and technicians. Further research is needed to assess the economic benefits of CE approaches across different industries and locations [18]. Scaling up and replicating successful circular practices in different contexts also requires more investigation (as highlighted in reports by the Ellen MacArthur Foundation) [19, 20]. Understanding the scalability and replicability of circular practices can help identify the most effective approaches and encourage wider adoption.

4 Findings

4.1 Circular Economy as a Concept

A closed-loop system is the foundation of the CE, which aims to cut down on material cycles and resource consumption [18]. CE focuses on procedures like manufacturing, design, recycling, and repair to reduce the amount of waste generated [11]. Sustainable industrial systems have intricate effects, so CE seeks to promote them at the micro, meso, and macro levels. The principles of "Resource-Product-Renewed resource" flow and "Reduction, Recycling, and Reuse"are all involved in the change from a linear system to a circular system [21].

Numerous studies that led to the discovery of CE in various applications can be seen in recent literature. Examples of recent studies on the supply chain, services, building industry, technology, and manufacturing are available. The CE has many advantages, including lower carbon emissions, better resource utilization, and increased business competitiveness [22]. But putting it into practice is challenging. One of the most significant issues is the need for more knowledge and understanding of how the CE operates. The development of the necessary infrastructure and policies is slowed by the lack of experience, which prevents buy-in and support from stakeholders (Fig. 16.1).

Moving from the current linear economy model, which is based on "take, make, and dump," to a CE model is challenging. Because the linear model has become so ingrained in business procedures, we require significant systems, behavior, and mindset adjustments. For instance, recycled materials might not be valued as highly as new materials or might not be required to be used in products. The current product manufacturing process also disregards end-of-life issues, which makes recycling

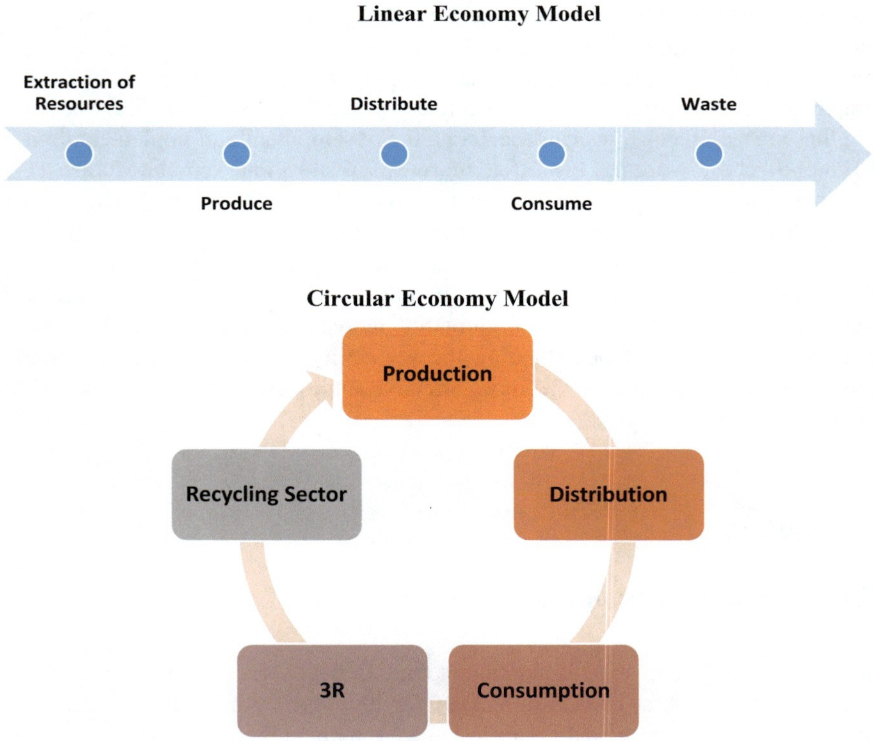

Fig. 16.1 The linear and circular economy model. (Source: Authors' own work)

and reuse challenging or even impossible [16]. Considerable changes are needed for the entire product lifecycle to move toward circular product design. Due to the absence of encouraging laws and regulations that support the adoption of CE, businesses find it challenging to defend the costs and investments necessary for adopting CE practices. Circular practices are less appealing unless obvious financial benefits exist because many businesses prioritize financial gains [1]. CE principles can be promoted by implementing tax breaks, subsidies, and other financial incentives.

4.2 SMEs Significance to the Economy

SMEs play a crucial role in the economy by facilitating innovation, generating employment prospects, and promoting economic expansion. SMEs play a significant role in the global business landscape. They employ between 50% and 60% of the workforce, a significant portion of the world's population. The number of SMEs is estimated to be around 42.50 million, with many of them (95%) being industrial units [23]. These SMEs are a crucial source of employment, providing opportunities for around 106 million individuals, constituting nearly 40% of the country's workforce.

Micro, small, and medium enterprises (MSME) comprise over 95% of industrial establishments and significantly contribute to the country's gross domestic product (GDP) growth. The target set by India is to achieve a renewable energy capacity of 450 Gigawatts by the year 2030; as per a study conducted by the Reserve Bank of India, the SMEs in India have provided employment opportunities to a staggering number of 117 million individuals, with a notable 20% of the workforce being situated in rural regions [24]. Micro, small, and medium enterprises (MSMEs) have a significant positive influence on India's gross domestic product (GDP) and generate a substantial number of employment opportunities, particularly in rural regions [23]. The Indian government has implemented various policies and programs to support micro, small, and medium enterprises [24]. These initiatives facilitate entrepreneurship, enhance skill development, and provide access to markets, technology, and financial resources. The SMEs in India's manufacturing and service sectors significantly contribute to the gross domestic product (GDP), with the manufacturing sector accounting for roughly 6.11% and the service sector for roughly 24.6% [25].

4.3 Challenges and Opportunities for SMEs to Adopt and Implement Circular Economy

SMEs frequently need help with the adoption and execution of CE practices. The challenges encountered in adopting sustainable practices may arise from insufficient resources, financial limitations, inadequate understanding and awareness of the CE principles, and the perceived intricacy of the transition process [16].

Adopting a CE necessitates substantial modifications in commercial frameworks, logistical networks, technological ailments, and functional procedures, presenting challenges for SMEs with restricted resources and established conventions [5]. Moreover, SMEs may need help modifying their existing business models and surmounting opposition to change from various stakeholders. Challenges in implementing the CE differ between developed and developing countries due to unique local conditions [26]. Factors influencing a country's position and other challenges can result in different barriers to development. While the CE offers multiple benefits, including carbon emission reduction, efficient resource utilization, and enhanced competitiveness of businesses, its implementation is not without difficulties. Some perceive it as too complex or costly to adopt, impeding stakeholder buy-in and hindering the establishment of necessary infrastructure and policies to support circular practices [27]. Notwithstanding these obstacles, adopting CE can provide many advantages to these enterprises.

Implementing CE practices has the potential to improve sustainability by decreasing waste, optimizing the utilization of resources, and minimizing environmental effects. In addition, incorporating CE principles can enhance customer engagement and bolster organizational image, given the growing consumer preference for environmentally responsible enterprises. Furthermore, it presents a prospect for SMEs to expand their technical expertise and investigate inventive approaches toward achieving resource efficiency. SMEs need access to educational and awareness initiatives, monetary backing, capacity enhancement programs, and customized counsel to assist them in navigating the complexities of executing CE practices proficiently. Recent literature has highlighted numerous studies that have explored the applications of CE. Examples of these studies encompass various sectors such as supply chain [28], services [29], construction [2], and manufacturing [4].

4.4 Significance of Blockchain Adoption for SME Businesses

Utilizing BC technology presents a considerable opportunity for enterprises, affording them advantages such as heightened transparency, bolstered security, and streamlined tracking functionalities [3]. Organizations have the potential to utilize this technology to monitor transactions, reclaim and reintegrate underutilized items, and enhance their sustainability initiatives [11]. The employment of BC technology has been progressively on the rise in the manufacturing industry, specifically to facilitate transparent transactions and advance sustainability efforts. The adoption of BC technology by enterprises can facilitate the exploration of novel avenues for enhancing operational efficacy, fostering trust, and promoting environmental sustainability [16].

BC technology can benefit small- and medium-sized enterprises (SMEs), allowing them to compete with larger organizations and improve efficiency. Several use cases already exist for implementation exclusively by SMEs too. Here are some key ways BC can be useful for SMEs.

BC technology offers numerous benefits to small- and medium-sized enterprises (SMEs) in the context of CE practices. It can revolutionize supply chain management by enabling real-time tracking and tracing of goods, ensuring transparency, accountability, and authenticity [20]. This is particularly valuable in industries such as pharmaceuticals, luxury goods, and automobile parts, where product integrity is crucial. The use of BC technology in supply chains can reduce the risk of counterfeit goods, improve quality assurance, and facilitate easier recall processes [30]. It provides SMEs with a reliable and efficient way to monitor inventory, streamline logistics, and enhance overall supply chain operations [7].

The BC technology enhances various financial aspects for SMEs. It facilitates the use of smart contracts, which are self-executing contracts with terms directly written into lines of code. These contracts automatically enforce obligations and agreements, reducing the need for intermediaries and saving time and money [31]. BC can also facilitate faster, cheaper, and more secure cross-border payments, which is particularly advantageous for SMEs engaged in international business [3]. By leveraging BC for payments and transfers, SMEs can benefit from improved cash flow, reduced transaction costs, and enhanced security. Additionally, BC can serve as a secure and immutable record-keeping system for SMEs, enabling them to maintain financial records, contracts, and other important documentation in a tamper-proof and easily auditable manner. This helps prevent fraud, simplifies auditing processes, and ensures data integrity for critical business information [20].

4.5 *Challenges in the Adoption of Blockchain*

There are problems with using BC technology in the CE. India needs to research how BC and the CE fit together more than other countries. To solve these problems, businesses and universities must collaborate and form partnerships with other companies to look for opportunities and develop solutions. By working together, stakeholders can promote CE practices, set up the necessary infrastructure, policies, and incentives, and make it easier for people to start using circular practices [27].

Studying the current state of CE adoption in small industries in India will show how they see things and give ideas on improving economic, environmental, and social performance. These projects can turn waste into something useful and help people learn more about using the CE [26]. Some companies have made significant progress toward zero waste, leading others to do the same [32].

Putting BC technology into the CE can have a lot of benefits, such as making it easier to track and see where things come from, speeding up transactions, and making it easier for people to trust each other. But problems like technical complexity, interoperability, scalability, data privacy, and regulatory frameworks need to be fixed before the full potential of BC can be used in CE projects [33]. Collaboration, thorough research, and a supportive ecosystem can pave the way for the successful adoption of blockchain-based CE practices, making it easier to move toward a more sustainable and efficient economic model [2].

4.6 Use of Blockchain in SMEs to Improve the Adoption of Circular Economy

The integration of BC technology offers a promising prospect for SMEs at the intersection of the CE. Integrating BC technology can allow MSMEs to improve transaction monitoring, resource management, and supply chain transparency and traceability [3]. The utilization of technology allows for the seamless tracking of metrics and the tracing of customers, streamlining the process of recovering and reintroducing surplus goods back into the market [34].

BC can make the supply chain more transparent by letting SMEs track where materials come from, check certifications, and make sure they are ethically sourced. BC also makes it possible to create digital records that can be used to prove the origin and authenticity of a product. This gives customers a lot of information about the materials used, the manufacturing process, and the product's lifecycle [35]. BC can also help keep track of and manage assets, making them more useful and making it easier to use them again or fix them up [16]. It can also simplify waste management and recycling by making tracking and verifying waste streams easy. Blockchain-based smart contracts can automate transactions like resource sharing, renting, and material exchange [36]. On BC platforms, SMEs can connect through collaborative networks and marketplaces, which helps build a CE ecosystem.

This makes it easier to adopt environmentally friendly practices and maximize the use of resources. Developing interest and change acceptance among SMEs is necessary to practically apply BC technology and CE principles. An open and receptive cognitive mindset is necessary for exploring new opportunities and having the flexibility to adapt and advance.

5 Discussion

The success of the CE heavily depends on collaboration and cooperation between various groups. Businesses, consumers, governments, and other stakeholders must collaborate to remove obstacles to adoption. To encourage the adoption of CE and its principles, this collaboration entails the development of the necessary infrastructure, regulations, and incentives [36]. Studies on adopting CE practices in Indian SMEs can provide information on the current state and potential advantages [11, 28]. Companies can improve their economic, environmental, and social performance by changing how they view waste and recognizing its value. Achieving zero waste and realizing the potential of CE can be accomplished by raising knowledge levels and incorporating CE practices into business operations [15]. The CE is a promising way to make SMEs more sustainable and self-renewing. Innovation is also important, and many examples exist of companies creating new business models and technologies to support circularity. Governments can do much through policy and regulation to help the CE grow [11]. For example, the European Union has adopted a Circular Economy Action Plan [37], which includes steps to promote the use of recycled materials, reduce waste, and encourage sustainable production and consumption. It is also clearly understood that the policy implications need to be revised and reimplemented through strict actions [38].

The study revealed that the public and officials involved in sustainable development lack awareness and lagged in understanding the basic principles of the CE. This needs to be addressed by the current researchers using a consultancy process, and educating the prime members can show impacts on SMEs. There are several risks associated with implementation in all sectors by the entire country [39].

Some of the barriers that companies lack in adoption are as follows: (a) capital is the most influential roadblock [40]; (b) there is a massive gap in the supply and demand network support [39]; (c) advanced technology is one of the constraints associated with company's capital [41]; and (d) the lack of cooperation from the government and legislative councils on the policy revisions and implications [42]. Cost is one thing, but changing people's mindsets is a challenge for implementing the CE entirely. That is precisely what is required to transform CE from a concept to a tangible reality for manufacturers and societies across the globe. BC technology can significantly aid SMEs in adopting CE practices.

BC enables SMEs to track and verify their products' origin, lifecycle, and sustainability credentials by providing transparency, traceability, and accountability. This ensures the ethical sourcing of materials, promotes resource efficiency, and encourages reuse, refurbishment, and recycling [1]. In addition, BC enables secure and efficient peer-to-peer transactions, encourages collaboration within CE networks, and streamlines processes such as supply chain management, waste management, and energy trading. BC enables small enterprises to adopt CE principles, reduce waste, improve sustainability, and contribute to a more circular and sustainable future [9]. This may enable SMEs to become more competitive, cost effective in the long run, and improve their sustainability.

6 Conclusion

The effective implementation of the circular economy (CE) requires collaboration among diverse stakeholders, including enterprises, consumers, governmental bodies, and other relevant entities. By overcoming obstacles and fostering collaboration, CE practices can be successfully adopted, leading to beneficial outcomes for the economy, environment, and society. Research on CE adoption among small- and medium-sized enterprises (SMEs) in India provides valuable insights. It emphasizes the importance of raising awareness and changing mindsets. Addressing various barriers, such as financial constraints, supply chain network inadequacies, technological limitations, and limited government collaboration, is crucial. In this context, the utilization of blockchain technology holds promise for enhancing transparency and efficiency in circular economy practices. By embracing circular economy principles and leveraging technological advancements, SMEs have the potential to enhance their competitiveness, cost-efficiency, and sustainability, thus contributing to the realization of a circular and sustainable future.

Declaration of Conflicting Interest According to the authors, no conflicts of interest existed with this paper's research, writing, or publishing.

References

1. Lin, K. Y. (2018). User experience-based product design for smart production to empower industry 4.0 in the glass recycling circular economy. *Computers & Industrial Engineering, 125*, 729–738.
2. Sadeghi, M., Mahmoudi, A., Deng, X., & Luo, X. (2023). Prioritizing requirements for implementing blockchain technology in construction supply chain based on circular economy: Fuzzy Ordinal Priority Approach. *International journal of Environmental Science and Technology, 20*(5), 4991–5012.
3. Godase, N., Rajasekhar, M., Kondala, M., Yadav, P., & Agarwal, M. (2023). Blockchain-based supply chain management (BC-SCM): Challenges. *Benefits & Solutions. European Economic Letters (EEL), 13*(3), 141–149.
4. Lieder, M., & Rashid, A. (2016). Towards circular economy implementation: A comprehensive review in context of manufacturing industry. *Journal of Cleaner Production, 115*, 36–51.
5. de Sousa, L., Jabbour, A. B., Jabbour, C. J. C., Godinho Filho, M., & Roubaud, D. (2018). Industry 4.0 and the circular economy: A proposed research agenda and original roadmap for sustainable operations. *Annals of Operations Research, 270*, 273–286.
6. Centobelli, P., Cerchione, R., Chiaroni, D., Del Vecchio, P., & Urbinati, A. (2020). Designing business models in circular economy: A systematic literature review and research agenda. *Business Strategy and the Environment, 29*(4), 1734–1749.
7. Nayal, K., Kumar, S., Raut, R. D., Queiroz, M. M., Priyadarshinee, P., & Narkhede, B. E. (2022). Supply chain firm performance in circular economy and digital era to achieve sustainable development goals. *Business Strategy and the Environment, 31*(3), 1058–1073.
8. Kitchenham, B., Brereton, O. P., Budgen, D., Turner, M., Bailey, J., & Linkman, S. (2009). Systematic literature reviews in software engineering–a systematic literature review. *Information and Software Technology, 51*(1), 7–15.
9. Kouhizadeh, M., Zhu, Q., & Sarkis, J. (2020). Blockchain and the circular economy: Potential tensions and critical reflections from practice. *Production Planning & Control, 31*(11–12), 950–966.
10. Xue, B., Chen, X. P., Geng, Y., Guo, X. J., Lu, C. P., Zhang, Z. L., & Lu, C. Y. (2010). Survey of officials' awareness on circular economy development in China: Based on municipal and county level. *Resources, Conservation and Recycling, 54*(12), 1296–1302.
11. Nudurupati, S. S., Budhwar, P., Pappu, R. P., Chowdhury, S., Kondala, M., Chakraborty, A., & Ghosh, S. K. (2022). Transforming sustainability of Indian small and medium-sized enterprises through circular economy adoption. *Journal of Business Research, 149*, 250–269.
12. Gunasekaran, A., Subramanian, N., & Rahman, S. (2015). Green supply chain collaboration and incentives: Current trends and future directions. *Transportation Research Part E: Logistics and Transportation Review, 74*, 1–10.
13. Tranfield, D., Denyer, D., & Smart, P. (2003). Towards a methodology for developing evidence-informed management knowledge by means of systematic review. *British Journal of Management, 14*(3), 207–222. https://doi.org/10.1111/1467-8551.00375
14. Glock, C. H., Grosse, E. H., & Ries, J. M. (2014). The lot sizing problem: A tertiary study. *International Journal of Production Economics, 155*, 39–51.
15. Nunes, B., Batista, L., Masi, D., & Bennett, D. (2022). *Sustainable operations management: Key practices and cases*. Taylor & Francis.

16. Gong, Y., Xie, S., Arunachalam, D., Duan, J., & Luo, J. (2022). Blockchain-based recycling and its impact on recycling performance: A network theory perspective. *Business Strategy and the Environment, 31*(8), 3717–3741.
17. Jovell, D., Pou, J. O., Llovell, F., & Gonzalez-Olmos, R. (2021). Life cycle assessment of the separation and recycling of fluorinated gases using ionic liquids in a circular economy framework. *ACS Sustainable Chemistry & Engineering, 10*(1), 71–80.
18. Flores, L., Josa, I., García, J., Pena, R., & Garfí, M. (2023). Constructed wetlands for winery wastewater treatment: A review on the technical, environmental and socio-economic benefits. *Science of the Total Environment, 882*, 163547.
19. MacArthur, E., Zumwinkel, K., & Stuchtey, M. R. (2015). *Growth within: A circular economy vision for a competitive Europe*. Ellen MacArthur Foundation.
20. Upadhyay, A., Mukhuty, S., Kumar, V., & Kazancoglu, Y. (2021). Blockchain technology and the circular economy: Implications for sustainability and social responsibility. *Journal of Cleaner Production, 293*, 126130.
21. Reike, D., Vermeulen, W. J., & Witjes, S. (2018). The circular economy: New or refurbished as CE 3.0?—Exploring controversies in the conceptualization of the circular economy through a focus on history and resource value retention options. *Resources, Conservation and Recycling, 135*, 246–264.
22. Natarajan, S., Akshay, M., & Aravindan, V. (2022). Recycling/reuse of current collectors from spent lithium-ion batteries: Benefits and issues. *Advanced Sustainable Systems, 6*(3), 2100432.
23. Ministry of MSME. (2021). *Annual report 2021/22*. Government of India. [Online]. Available at: https://msme.gov.in/sites/default/files/MSMEENGLISHANNUALREP ORT2021-22.pdf. Accessed 28 Apr 2023
24. Ministry of MSME. (2017). *Annual report 2016/17*. Government of India. [Online]. Available at: https://msme.gov.in/sites/default/files/MSME%20ANNUAL%20REPORT%20 201617%20ENGLISH.pdf. Accessed 15 May 2023
25. Chakrabarty, A., Norbu, T., & Mall, M. (2020). Fourth industrial revolution: Progression, scope and preparedness in India—Intervention of MSMEs. In *Intelligent computing in engineering: Select proceedings of RICE 2019* (pp. 221–228).
26. Ghisellini, P., Cialani, C., & Ulgiati, S. (2016). A review on circular economy: The expected transition to a balanced interplay of environmental and economic systems. *Journal of Cleaner Production, 114*, 11–32.
27. Idrees, S., & Nowostawski, M. (2022). *Transformations through Blockchain technology*. Springer.
28. Mastos, T. D., Nizamis, A., Terzi, S., Gkortzis, D., Papadopoulos, A., Tsagkalidis, N., Ioannidis, D., Votis, K., & Tzovaras, D. (2021). Introducing an application of an industry 4.0 solution for circular supply chain management. *Journal of Cleaner Production, 300*, 126886.
29. Pishdar, M., Danesh Shakib, M., Antucheviciene, J., & Vilkonis, A. (2021). Interval type-2 fuzzy super sbm network deal for assessing sustainability performance of third-party logistics service providers considering circular economy strategies in the era of industry 4.0. *Sustainability, 13*(11), 6497.
30. Kumar, M. G. V., Chande, K., Kanekar, R., Kondala, M., Majid, M. A. A., & Patil, P. P. (2022, November). The role of block chain integration in the field of food supply chain in the present and future development. In *2022 international interdisciplinary humanitarian conference for sustainability (IIHC)* (pp. 1543–1549). IEEE.
31. Yang, L., Zou, H., Shang, C., Ye, X., & Rani, P. (2023). Adoption of information and digital technologies for sustainable smart manufacturing systems for industry 4.0 in small, medium, and micro enterprises (SMMEs). *Technological Forecasting and Social Change, 188*, 122308.
32. Yu, Z., Khan, S. A. R., Ponce, P., Muhammad Zia-ul-haq, H., & Ponce, K. (2022). Exploring essential factors to improve waste-to-resource recovery: A roadmap towards sustainability. *Journal of Cleaner Production, 350*, 131305.
33. Xie, S., Gong, Y., Kunc, M., Wen, Z., & Brown. (2022). The application of blockchain technology in the recycling chain: A state-of-the-art literature review and conceptual frame-

work. *International Journal of Production Research*, 1–27. https://doi.org/10.1080/00207543.2022.2152506
34. Jain, G., Kamble, S. S., Ndubisi, N. O., Shrivastava, A., Belhadi, A., & Venkatesh, M. (2022). Antecedents of Blockchain-Enabled E-commerce Platforms (BEEP) adoption by customers–A study of second-hand small and medium apparel retailers. *Journal of Business Research, 149*, 576–588.
35. Soldatos, J., Kefalakis, N., Despotopoulou, A. M., Bodin, U., Musumeci, A., Scandura, A., Aliprandi, C., Arabsolgar, D., & Colledani, M. (2021). A digital platform for cross-sector collaborative value networks in the circular economy. *Procedia Manufacturing, 54*, 64–69.
36. Mazumdar, S. (2023). How to reduce information silos while Blockchain-ifying recycling focused supply chain solutions? In *The 56th Hawaii international conference on system sciences. HICSS 2023* (pp. 459–468). Hawaii International Conference on System Sciences (HICSS).
37. European Commission. (2011). *Communication from the commission to the European Parliament, the council, the European economic and social committee and the Committee of the Regions Youth Opportunities Initiative*. European Commission.
38. Geng, Y., & Doberstein, B. (2008). Developing the circular economy in China: Challenges and opportunities for achieving 'leapfrog development'. *The International Journal of Sustainable Development & World Ecology, 15*(3), 231–239.
39. Ding, S., Tukker, A., & Ward, H. (2023). Opportunities and risks of internet of things (IoT) technologies for circular business models: A literature review. *Journal of Environmental Management, 336*, 117662.
40. Rizos, V., Behrens, A., Van der Gaast, W., Hofman, E., Ioannou, A., Kafyeke, T., Flamos, A., Rinaldi, R., Papadelis, S., Hirschnitz-Garbers, M., & Topi, C. (2016). Implementation of circular economy business models by small and medium-sized enterprises (SMEs): Barriers and enablers. *Sustainability, 8*(11), 1212.
41. Oliva, G., Galang, M. G., Zarra, T., Belgiorno, V., & Naddeo, V. (2023). Advanced technologies for a smart and integrated control of odour emissions from wastewater treatment plant. In *Current developments in biotechnology and bioengineering* (pp. 315–332).
42. Razminiene, K. (2019). Circular economy in clusters' performance evaluation. *Equilibrium. Quarterly Journal of Economics and Economic Policy, 14*(3), 537–559.

Index

S. M. Idrees, M. Nowostawski (eds.), *Blockchain Transformations*, Signals and Communication Technology, https://doi.org/10.1007/978-3-031-49593-9

Printed in the United States
by Baker & Taylor Publisher Services